ns to Microcirculation

BIOPHYSICS AND BIOENGINEERING SERIES

A Series of Monographs and Edited Treatises

EDITOR

ABRAHAM NOORDERGRAAF

Department of Bioengineering D2
University of Pennsylvania
Philadelphia, Pennsylvania

1. ABRAHAM NOORDERGRAAF. Circulatory System Dynamics. 1978
2. MARY PURCELL WIEDEMAN, RONALD TUMA, AND HARVEY NORMAN. An Introduction to Microcirculation

In preparation

3. ICHIJI TASAKI. Physiology and Electrochemistry of Nerve Fibers

An Introduction to Microcirculation

MARY PURCELL WIEDEMAN
RONALD F. TUMA

Department of Physiology
Temple University School of Medicine
Philadelphia, Pennsylvania

HARVEY NORMAN MAYROVITZ

Research Division
Miami Heart Institute
Miami Beach, Florida
and
Department of Physiology
Temple University School of Medicine
Philadelphia, Pennsylvania

1981

ACADEMIC PRESS
A Subsidiary of Harcourt Brace Jovanovich, Publishers
New York London Toronto Sydney San Francisco

COPYRIGHT © 1981, BY ACADEMIC PRESS, INC.
ALL RIGHTS RESERVED.
NO PART OF THIS PUBLICATION MAY BE REPRODUCED OR
TRANSMITTED IN ANY FORM OR BY ANY MEANS, ELECTRONIC
OR MECHANICAL, INCLUDING PHOTOCOPY, RECORDING, OR ANY
INFORMATION STORAGE AND RETRIEVAL SYSTEM, WITHOUT
PERMISSION IN WRITING FROM THE PUBLISHER.

ACADEMIC PRESS, INC.
111 Fifth Avenue, New York, New York 10003

United Kingdom Edition published by
ACADEMIC PRESS, INC. (LONDON) LTD.
24/28 Oval Road, London NW1 7DX

Library of Congress Cataloging in Publication Data

Wiedeman, Mary P.
  An introduction to microcirculation.

  (Biophysics and bioengineering; 2)
  Includes bibliographies and index.
  1. Microcirculation. I. Tuma, Ronald Franklin.
II. Mayrovitz, Harvey Norman. III. Title. IV. Series.
[DNLM: 1. Microcirculation. WL BI878 v. 2 / WG 104
W644i]
QP106.6.W52   612'.135   81-3521
ISBN 0-12-749350-6   AACR2

PRINTED IN THE UNITED STATES OF AMERICA

81 82 83 84   9 8 7 6 5 4 3 2 1

The authors of this book represent two generations of students of microcirculation, the first generation originating with Professor Paul A. Nicoll, to whom this book is affectionately dedicated.

Contents

Preface *xi*

INTRODUCTION

1 Historical Introduction

Text *3*
References *11*

2 General Anatomical Comparisons

Introduction *12*
References *20*

3 Microvasculature of Specific Organs and Tissues

- I. Muscle   21
- II. Visceral Organs   33
- III. Lung   51
- IV. The Bulbar Conjunctiva   53
- V. Pia Mater   55
- VI. Special Tissues   57
- VII. Conclusion   70
-     References   70

4 Methods of Preparation of Tissues for Microscopic Observation

- I. Introduction   75
- II. Exteriorization of Internal Tissues   77
- III. Transparent Chambers   82
- IV. *In Situ* Tissues and Organs   85
- V. Superficial Structures   91
- VI. Conclusion   94
-     References   94

REGULATION OF FLOW AND EXCHANGE

5 Factors Involved in the Regulation of Blood Flow

- I. Introduction   99
- II. Control of Microcirculation   100
- III. Control of Skeletal Muscle Circulation   112
- IV. Control of Cerebral Circulation   119
- V. Control of Cardiac Blood Flow   122
- VI. Control of Gastrointestinal Circulation   123
- VII. Control of Cutaneous Circulation   129
- VIII. Autoregulation   131
- IX. Summary   135
-     References   137

6 Exchange in the Microcirculation

- I. Introduction *140*
- II. Diffusion *142*
- III. Filtration and Osmosis *145*
- IV. Vesicular Transport *147*
- V. Summary *151*
- References *152*

HEMODYNAMICS

7 Quantitative Techniques for Measurement of Velocity and Pressure of Blood

- I. Measurements of Blood Velocity in the Microcirculation *157*
- II. Measurement of Blood Pressure in the Microcirculation *169*
- References *174*

8 Hemodynamics of the Microcirculation

- I. The Conceptual Framework *177*
- II. Microcirculatory Pressure and Flow *in Vivo* *182*
- III. The Significance of Design, Cells, and Capillary Flow *191*
- IV. Other Factors Influencing *in Vivo* Hemodynamics *211*
- References *213*

Index *217*

Preface

There is great need for a book containing the most basic information about the microcirculation, as we have recognized through our own past experiences in the laboratory and in dealing with the myriad questions from medical students, graduate students, clinicians, and young investigators. There are very few medical textbooks which contain descriptive material of the terminal vascular beds beyond an oversimplified diagram, fewer which contain discussions of regulation of blood flow in this area, and essentially none which are in complete agreement regarding structure, nomenclature, or control.

It is not our intent to present rigid definitions for the structural and functional aspects of the microvasculature, but rather to consolidate current information gained from the numerous vascular beds that have been used for *in vivo* microscopic observations, to note the similarities and differences in architecture and function, to reveal the origin of certain terms and concepts, and to discuss hemodynamics of the microvessels.

This selection of material should enable an interested scientist to learn the essential features of the microcirculation. With this source of background information, one may develop a better understanding of current research in order to pursue his own more effectively.

More detailed information can be found in books on specific aspects and in investigative work in journals. This book is intended to be a primer of essential basic information.

Mary P. Wiedeman

Ronald F. Tuma

Harvey N. Mayrovitz

INTRODUCTION

1

Historical Introduction

William Harvey (1578–1657), eventually appointed private physician to Charles I of England, risked achieving this enviable position by persisting in the postulation that blood flowed in a circle from the heart, through the arteries, into the veins, and back to the heart. This postulate was contrary to the teachings of Galen, whose ideas and tenets had dominated medicine for the preceding 1200 years, and to suggest any deviation from the long standing theories of Galen was considered heresy. Harvey's conviction of the true nature of movement of blood through vessels, and his dedication to demonstrating it obscured the dangers inherent in the continuation of these studies and in the presentation of his findings. Reviews and analyses of his famous treaty, *De moto cordis,* published in 1628, are numerous and will not be repeated here, but mention must be made of his strongest argument in support of his thesis. Chapter 9 of the 72-page work is entitled "The circulation of the blood is proved by a prime consideration." It is here that Harvey explains why blood ejected from the heart must be recirculated. If, he reasons, the ventricle of the heart contains some small quantity of blood, perhaps an ounce or an ounce and a half, and if some small quantity of that small amount is ejected with each systole, then in the course of one-half hour in which the heart beats more than 1000 times, from ten to eighty pounds of blood would be ejected, which is much more than is contained in the entire body. Therefore, the blood must move from the heart and back again to be repeatedly ejected during systole.

1. Historical Introduction

There is no evidence that Harvey looked for the connection between the arterial and venous vessels, which he termed porosities of the flesh, nor did he live long enough to know of the discovery of the small, hairlike (capillary) vessels that joined artery with vein by Malpighi in 1661.

Dr. Harvey lived 79 years. He is buried in a small parish church in Hempstead, Essex not far from Caius College in Cambridge, where he had been admitted at the age of 16 (Fig. 1.1.)

Harvey's life spanned an exciting period of history, and he was exposed to the prominent figures of the time. He was associated with Galileo when he studied in Padua; he was known to Elizabeth I, whose encouragement of free thinking and whose efforts to secure the release from restrictive religious laws benefited him; he attended two Kings of England, James I and Charles I; and the importance of his concept that blood went in a circle from the heart and back was recognized during his lifetime (Fig. 1.2).

The existence of pathways that made circulation possible was confirmed by the microscopic observations of Marcello Malpighi, a professor of medicine in Bologna (Fig. 1.3). Malpighi described blood vessels and the flow of blood,

Fig. 1.1. St. Andrews Church, Hempstead, Essex where Sir William Harvey is buried.

Historical Introduction

Fig. 1.2. A popular portrait of Sir William Harvey, painted in 1626.

which he observed in the lungs of frogs. Malpighi was not looking for confirmation of Harvey's postulate, but had been engaged in studies at a place and a time when the use of magnifying lenses to enlarge the image of almost anything was a popular pursuit.

Marcello Malpighi was born in 1628 near Bologna to parents apparently in comfortable circumstances. The century in which he was born has been described as a bigoted, superstitious, and cruel age. The universities then were mostly concerned with preserving old information. It was an age of plagues and almost

"I never reached my idea...by the aid of books, but by the long, patient and varied use of the microscope".

Malpighi, 1666

Fig. 1.3. Marcello Malpighi. This portrait is in The Academy of Fine Arts in Bologna.

Historical Introduction

Fig. 1.4. The tomb of Malpighi in the Church of San Gregorio, Bologna.

continuous wars in Europe and the restraints imposed by religious beliefs were tremendous; the Catholic lands were dominated by the dogmas of the church, the Protestant ones by the authority of the bible. Malpighi spent most of his life in Bologna, as a student and as a practitioner of medicine. He died in 1694, and his remains can be found in the small church of La Chiesa di San Gregorio in Bologna. (See Fig. 1.4).

In 1300, "eye glasses" or "spectacles" were already in use; thus, the magnifying effect of lenses was known even at that early period. Telescopes were popular instruments at the end of the sixteenth century, and it has been suggested (Bradbury, 1967) that the idea of the microscope may have originated with the opticians in Holland, who were so expert in making lenses for spectacles and telescopes. The use of lenses was highly developed in Italy at this time. It was known that lenses had the ability to extend the utility of the human eye. Indeed, the development of the microscope was pursued with great vigor in Italy, and major advances were made by Campani and Divina, both well known for their skills in the manufacture of microscopes in the middle 1600s. The type of microscope probably used by Malpighi is shown in Fig. 1.5.

Fig. 1.5. An Italian microscope of the type used by Malpighi. (From Bradbury, 1967.)

Historical Introduction

In keeping with such rapid development, it was not long before a microscope designed especially for examining the circulation of blood was introduced by John Marshall in England, complete with a lead coffin for restraining the small wiggling fish under observation (Fig. 1.6). Also during this time, an ingenious man, Antoni van Leeuwenhoek, was not only making microscopes, but was making amazing observations of blood flow in living animals. He wrote of the blood flow in the transparent tails of frog worms (tadpoles) as follows:

> So that it appeared to me that the blood vessels, observed in this animal and called arteries and veins, are exactly alike; only they can be called arteries when they carry blood to the utmost extremeties of the small blood vessels, and veins when they carry the same blood back to the heart.

Today, almost three centuries later, there is no better way to determine if one is observing an arterial or venous vessel in an *in vivo* microscopic preparation.

In the early 1700s an Anglican minister named Stephen Hales performed the miraculous feat of making the first direct measurement of blood pressure. He used a horse as his experimental animal, cannulated the carotid artery, which he connected to a glass tube by means of a goose trachea, and watched the blood rise

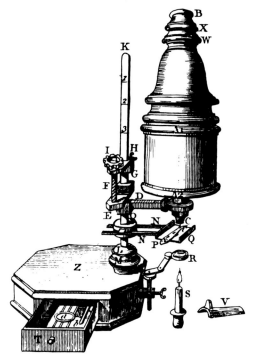

Fig. 1.6. The microscope designed by Marshall for observing circulation of the blood in a fish's tail. (From Bradbury, 1967.)

9 feet, 8 inches in the tube. The direct measurement of blood pressure in experimental animals is essentially the same today; the major difference is that we impose the weight of a column of mercury or a pressure equilibrated against the mercury column and so express blood pressure in terms of the height of a column of mercury in millimeters. Hales also described blood flow in very small blood vessels slightly larger than the size of a "globule" (i.e., red blood cell) and noted that these vessels alternately contracted and relaxed. About 25 years later, Robert Whytt discussed contraction of small arterial vessels as an aid to moving blood after the pulsations produced by the ejection of blood from the heart had diminished to the extent that there was not adequate force to push the blood forward. Haller in 1756 argued against contraction of small arterial vessels. Philip, in 1826, tried to demonstrate the effect of the central nervous system on blood flow while Black in 1825 supported the view that blood flow in the capillaries was independent of the heart and the central nervous system. Marshall Hall in 1831 attempted to define the components of terminal vascular beds and elaborated on van Leeuwenhoek's classification of vessels. Hall saw that capillary vessels differed from pre- and postcapillary vessels and called true capillaries those vessels that "do not become smaller by subdivision, nor larger by conjunction." Johannes Muller in 1838 advanced the descriptive material by measuring capillary diameters in numerous tissues, and he advanced the idea that there were invisible pores in capillary vessels that permitted fluid to escape.

In 1852, an interesting report by T. W. Jones appeared in which rhythmical contractility of veins was described and the function of valves seen in veins was discussed. Jones also noted that a "viscid-looking grayish granular lymph" was produced in veins following compression and that it could obstruct the vein completely before breaking off in small pieces. Jones was obviously describing platelet aggregation, although platelets and their role in hemostasis had not yet been discovered.

During the next seventy years, new information regarding structure, control, and behavior of microscopic blood vessels appeared in the literature. The state of the art was reviewed in 1922 by August Krogh, who gave a series of lectures in the United States shortly after he had received the Nobel prize in medicine and physiology. The material was published in the now classical book, *The anatomy and physiology of capillaries,* which contains the known and controversial material of that time (Krogh, 1929). In 1868, Stricher had described independent contractility of capillaries in frogs, and in 1873 Rouget saw cells on capillaries that contracted. This paper was forgotten for a time, but revived by Mayer in 1902 when he discovered branched cells on capillaries. As a result, the erroneous concept that capillaries were contractile developed. Subsequent research has dispelled this concept. In 1877, diapedesis of red blood cells was described by Cohnheim, and Hoyer described direct anastomosis between arteries. Spalteholz published his arrangement of blood vessels in striated muscle in 1888, a paper which is still referred to in the present day. A more detailed history of this can be found in Adelman, 1966; Chavois, 1975; and others.

The history of the development of knowledge of the capillary vessels is a topic in itself, and space does not permit more than just the citation of a few examples of the type of investigations that were carried out, but Krogh's book did act as a stimulus to future investigators who strove to refute or confirm the published facts.

The next two decades after Krogh were rich in imaginative and rewarding research. Scientists of this period have formed our present schools of thought; some of them are still productive, and their influence is reflected in current research.

REFERENCES

Adelman, H. B. (1966). "Marcello Malpighi and the Evolution of Embryology." Vol. 1, pp. 658-661, Cornell Univ. Press, Ithaca, New York.

Bradbury, S. (1967). "The Evolution of the Microscope." Pergamon, Oxford.

Chauvois, L. (1957). "William Harvey." Philos. Libr., New York.

Fishman, A. P., and Richards, D. W. (1964). "Circulation of the Blood, Men and Ideas." Oxford Univ. Press, London and New York.

Fulton, J. F. (1930). "Selected Readings in the History of Physiology." Thomas, Springfield, Illinois.

Krogh A. (1929). "The anatomy and physiology of capillaries." Yale Univ. Press, New Haven, Connecticut.

Wiedeman, M. P. (1974). "Microcirculation," Benchmark Papers in Human Physiology. Dowden, Hutchinson & Ross, Stroudsburg, Pennsylvania.

Willis, R. (1965). "The Works of William Harvey, M.D." Johnson Reprint, New York.

2

General Anatomical Comparisons

INTRODUCTION

The recycling of ideas and generalizations about microcirculation, its structure, its terminology, and its control continue with the result that certain concepts have become entrenched (Wiedeman, 1963, 1967, 1976; Zweifach, 1973, 1977). In a recent publication, Zweifach (1977) has restated several of these generalizations. (1) Although the tendency has been to emphasize differences rather than similarities, it is recognized that all circulatory beds share common features that must be identified if investigators are to understand one another. (2) Common features of vascular patterns are modified by the parenchymal structure of the various tissues in which the vessels lie. (3) The vessels of microcirculation cannot be identified on the basis of diameter, but should be classified from other characteristics, such as the physical location of a vessel in a branching sequence. (4) The cause of intermittent flow through capillaries is not the same in all tissues, but resides primarily in the varying contractile activity of precapillary sphincters or terminal arterioles.

A survey of the microvascular patterns presently known makes it obvious that although there are many similarities, it is not possible to present a "typical capillary bed" without numerous qualifying statements. In addition, a verbal description suitable for all precapillary, capillary, and postcapillary vessels cannot be written because of the variations found in different tissues and, in some

Introduction

cases, in different species that have been examined. It is necessary, however, to be aware of the features that are common to the majority of the various vascular beds and to find acceptable nomenclature for the vessels that are identified with the microcirculation. At this time there is enough descriptive material in the literature to enable us to recognize the common features of the microvascular beds in different tissues.

One approach to reach agreement about what is typical of most beds is to review the origin and development of current diagrams and descriptions that appear most frequently in reference books. Microvascular beds in specific tissues that have been explored or used as experimental sites will be reviewed later.

The most widely copied diagrams of a capillary bed are those of Zweifach that appeared in the literature in 1937, 1944, and 1949 (Zweifach, 1937; Chambers and Zweifach, 1944; Zweifach, 1949). Because of their ubiquity and the fact that they are usually the first diagrams to be seen as an introduction to the microcirculation it is important to review them first.

In 1937, Zweifach presented an extensive and detailed description of the capillary bed of the frog mesentery with several diagrams. The one most frequently used from this paper is seen in Fig. 2.1. In his description of this bed, Zweifach made the following points: (1) There are direct channels between the arterial and venous vessels that have a patent lumen at all times; they are continuations of arterioles and are called a-v bridges. The bridges have widely separated smooth muscle cells that are very thin and can be stimulated to contract. Because of the sparseness of their muscle cells, these vessels function as capillaries to supply the surrounding tissue with nutritive material. (2) The rest of the capillary

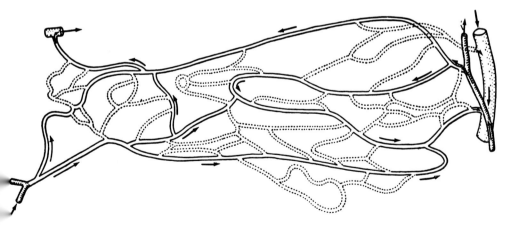

Fig. 2.1. Capillary bed of frog mesentery showing direct connections between arterial and venous vessels (a-v bridges). True capillaries are indicated by broken lines. (From Zweifach, 1937.)

bed is made up of nonmuscular true capillaries that are side branches of a-v bridges. The true capillaries provide a circulatory bed that is in use when an increased vascular surface is needed for exchange of substances for cellular nourishment.

In 1944, Chambers and Zweifach presented a diagram of a functional unit of a capillary bed (Fig. 2.2). The diagram was a composite of microscopic observations, primarily of rat mesoappendix, although dog omentum was found to be similar. Once again the idea of a thoroughfare vessel or central channel with capillaries arising from it dominates. The initial portion of the central channel, which is the contractile portion, is described as having muscle cells that are discontinuous, and this proximal muscular portion is called the metarteriole. As the vessel continues to the venous side, the muscle cells disappear. Precapillaries, the vessels that branch off of the metarteriole, are encircled with contrac-

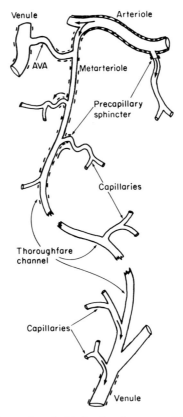

Fig. 2.2. Diagram of a capillary bed indicating the metarteriole and thoroughfare channel from which capillaries arise or enter the main pathway from arteriole to venule. AVA, arteriovenous anastomosis. (From Chambers and Zweifach, 1944.)

Introduction

tile muscle cells at their origins that are called precapillary sphincters. Farther along the central or thoroughfare channel, muscle cells at the branch sites disappear.

A central channel with its branches and their true capillaries is defined in this paper as a structural unit (= functional unit). Capillaries can communicate with capillaries of adjacent units, and a number of units make up a capillary bed. The 1944 diagram of the functional unit illustrates the components of a capillary bed as thoroughfare or central channels from which capillaries branch; the proximal contractile portion of this channel is called the metarteriole; precapillaries with precapillary sphincters; and wide capillaries at the venular end of the central channel.

In his paper of 1949, Zweifach introduced two new descriptive figures of the capillary bed and included two figures from his 1939 paper. One of the new figures was a camera lucida drawing of cat mesentery (Fig. 2.3) and the other was a schematic representation of the structural pattern of the capillary bed (Fig. 2.4). The second of the new figures was later mislabeled as a bed seen in muscle (Burton, 1965), and the error has been perpetuated by use of the diagrams from Burton's book by other authors. The two older diagrams from 1939 included the capillary bed as shown in Fig. 2.1 and the diagram of the capillary bed in the nail bed of human skin (Fig. 2.5).

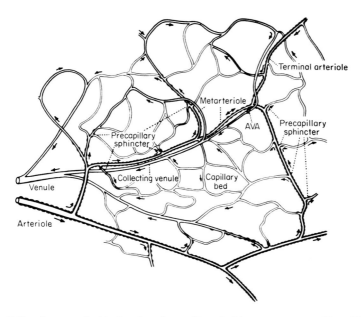

Fig. 2.3. A camera lucida drawing of a capillary bed in cat mesentery. (From Zweifach, 1949.)

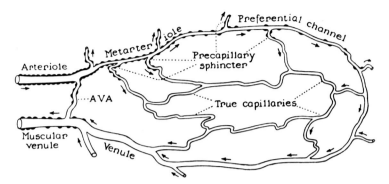

Fig. 2.4. A schematic representation of the structural pattern of the capillary bed. (From Zweifach, 1949.)

In *Fulton's Textbook of Physiology,* published in 1955, a modification of the functional unit shown in Fig. 2.2 appeared, in which connections were made between the various vascular components (Fig. 2.6A). These early diagrams continue to be used in the most current medical physiology textbooks and cardiovascular physiology monographs, and they are representative of most vascular beds that have been observed microscopically with one or two exceptions of specific terms. It is now recognized that the metarteriole as currently defined may not be a common feature of microvascular beds (Baez, 1977), and the precapillary sphincter may vary in position and in occurrence from tissue to tissue (Wiedeman, 1976).

In terminal vascular beds other than mesentery in which the metarteriole and preferential channel are prominent features, the components are usually presented as follows; from the arterial to the venous side, the vessels included in the

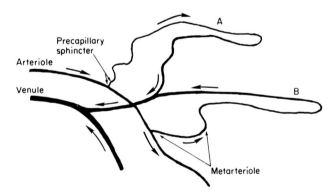

Fig. 2.5. Diagram to show capillary loops with a precapillary sphincter controlling flow through each loop. (From Zweifach, 1949.)

Introduction

microcirculation are the arteriole, the terminal arteriole, the capillary, the postcapillary venule, and the venule. Descriptions appropriate for the identification of these vessels are found below.

Arterioles of the microvasculature are parent vessels of the terminal arterioles. Terminal arterioles are so designated because they terminate in a capillary network without connection with any other arterial or venous vessel via an anastomosis or arcuate structure. The terminal arterioles are the site of the final muscular investment, which is called the precapillary sphincter. Beyond the precapillary sphincter the vessel continues as a pure endothelial tube, the capillary.

The definition of the capillary is the easiest because of the acceptance that it is a vessel, composed of endothelial cells, which lies between the distributing arterial vessels and the collecting venous vessels. The capillary has no encircling smooth muscle cells and, therefore, does not have the capacity to actively change diameter. It also has no need for vasoconstrictor nerves. Details of its ultrastructure that permit it to function as a primary site of exchange are well documented. Most capillaries originate from terminal arterioles as single vessels or as a burst of vascular pathways that form interlacing networks and end by converging to form a postcapillary venule. The exact site of the origin and termination of a capillary are difficult to identify by light microscopy. Indeed, the vessels show variable length, variable direction, which can be straight or circuituous before converging with another vessel, and variable though limited diameter. Also, our present inability to distinguish the absence of a muscular investment by direct visualization makes it virtually impossible to delineate the exact beginning and end of a capillary in the vascular bed of a living animal.

Conversely, the postcapillary vasculature is easy to visualize microscopically in the living animal. Venous vessels are collecting vessels; thus, by definition, the postcapillary venule is formed when convergence of two capillaries first occurs. The postcapillary venule is a nonmuscular vessel, slightly larger than the capillaries which form it. These postcapillary venules join to form venules where smooth muscle cells reappear, and valves are seen in some tissues. Venules, in turn, enter other vessels at Y-shaped junctions to form a vein or to empty into already formed veins.

Some of the commonly used terms must now be considered in greater detail, in an attempt to erase confusion arising from differences in definitions.

The metarteriole and the preferential channel are terms used by adherents to Zweifach's concept of the structural unit. In this concept, a single vessel—the metarteriole, which is encircled by smooth muscle fibers except at its most distal portion—transverses the tissue to empty into a venule. The portion without muscle is called the preferential channel and receives tributaries from the adjacent capillary net. The capillaries originate from the metarteriole with smooth muscle cells at each branch site. This muscular junction is called the precapillary sphincter. The metarteriole-preferential channel combination, forming a pathway between terminal arteriole and venule, is open for blood flow even in ischemic tissue, when the precapillary sphincters are contracted and no blood flows through the capillaries.

This vascular organization has not been identified in numerous tissues in which the location of the precapillary sphincter and the vessel designated as the terminal arteriole are different. A comparison of the two arrangements is seen in Figs. 2.6A and 2.6B. The arteriole in Fig. 2.6A has since been called "terminal arteriole" (Baez, 1977). This makes the arteriole and terminal arteriole the same vessel in the two diagrams, while the terminal arteriole of one is the metarteriole of the other. The spiral smooth muscle cells extend further along the smallest vessels of the arteriolar distribution in the bed that does not have a metarteriole and preferential channel. In the overall general pattern there is great similarity.

The precapillary sphincter is another structure that has been variously located along the arteriolar distribution. A recent publication (Wiedeman, 1976) reviewed the past and present use of the term and made the following points. (1) The term evolved from direct microscopic observation of the microvasculature and was later adopted by investigators using indirect measurements to study blood flow and capillary exchange. This resulted in a change of the term precapillary sphincter to simply a precapillary resistance area without a definitive location. (2) Reasons were given to support the belief that the precapillary sphincter's location dictates that it is involved primarily in local control of flow through capillaries and is not a factor in regulating peripheral resistance. The plea was made that the precapillary sphincter should be defined as the last smooth muscle cell on an arteriole or a terminal arteriole and its position not restricted to a branch orifice.

If the precapillary sphincter is to be the "gatekeeper" cell that controls blood flow through the capillaries, it should be located at the most distal site where muscular contraction, by closing the gate, shuts off flow to a capillary bed. To assign it to another location deprives it of the ultimate control of capillary flow. Zweifach (1977) notes that if control was located more proximally in the microcirculatory system, it would result in a shifting of flow in the capillary network as a whole, which would preclude active adjustments within a tissue. Such an arrangement would lead to a less efficient distribution of blood to discrete areas.

In summary, certain generalizations can be made regarding the architecture of the terminal arterial distribution, the capillary network, and the immediate postcapillary collecting vessels. The general pattern, as depicted by the most commonly used diagrams presented at the beginning of this chapter, are agreeable to the majority of scientists involved with the microcirculation. The minor difference relative to the vessel that is designated "terminal arteriole" by some and "metarteriole" by others, and also the true location of the precapillary sphincter must be recognized by investigators in order to reduce confusion in describing various activities at this level.

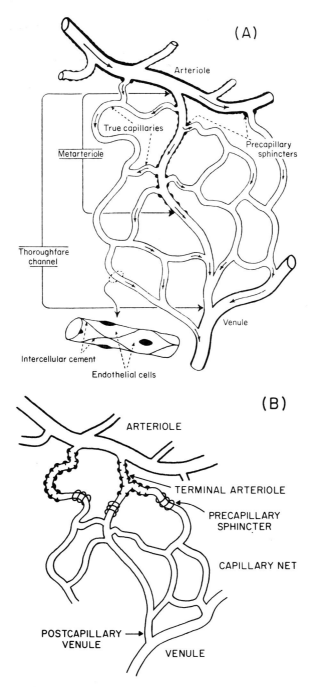

Fig. 2.6. (A) Schematic diagram of capillary bed according to Chambers and Zweifach and modified by Nelson. (From Fulton, 1955.) (B) Schematic diagram of a typical capillary bed showing different terminology for similar vessels.

Because of the variation in this general pattern in different tissues, specific characteristics of design imposed by both the structure and the function of the tissue must now be described.

REFERENCES

Baez, S. (1977). Microvascular terminology. *In* "Microcirculation" (G. Kaley and B. M. Altura, eds.) Vol. 1, pp. 23-35. Park Press, Baltimore, Maryland.

Burton, A. C. (1965). "Physiology and Biophysics of the Circulation." Year Book Publ., Chicago, Illinois.

Chambers, R., and Zweifach, B. W. (1944). The topography and function of the mesenteric capillary circulation. *Am. J. Anat.* **75,** 173-205.

Fulton, J. F. (1955). "Textbook of Physiology," 17th ed. Saunders, Philadelphia, Pennsylvania.

Wiedeman, M. P. (1963). Patterns of the arteriovenous pathways. *In* "Handbook of Physiology, Circulation II" (W. F. Hamilton and P. Dow, eds.) Sec. 2, Vol. II, pp. 891-933. Am. Physiol. Soc., Washington, D. C.

Wiedeman, M. P. (1967). Architecture of the terminal vascular bed. *In* "Physical Bases of Circulatory Transport: Regulation and Exchange" (E. B. Reeve and A. C. Guyton, eds.) pp. 307-312. Saunders, Philadelphia, Pennsylvania.

Wiedeman, M. P., Tuma, R. F., and Mayrovitz, H. N. (1976). Defining the Precapillary Sphincter. Microvasc. Res. 12: 71-75.

Zweifach, B. W. (1937). The structure and reactions of the small blood vessels in Amphibia. *Am. J. Anat.* **60,** 473-514.

Zweifach, B. W. (1949). Basic mechanisms in peripheral vascular homeostasis. *In* "Transactions of the Third Conference on Factors Regulating Blood Pressure" pp. 13-52. (B. W. Zweifach and E. Schorr, eds.) Josiah Macy, Jr. Found., New York.

Zweifach, B. W. (1973). Microcirculation. *Annu. Rev. Physiol.* **35,** 117-150.

Zweifach, B. W. (1977). Perspectives in microcirculation. *In* "Microcirculation" (G. Kaley and B. M. Altura, eds.) Vol. 1, pp. 1-21. Park Press, Baltimore, Maryland.

3

Microvasculature of Specific Organs and Tissues

I. MUSCLE

The descriptions of the vascular beds presented here are compiled from published material. Some tissues have been described in detail by investigators who have either developed the technique or used the beds for experimental purposes. Other tissues that have been used primarily for studies of behavioral characteristics of microvessels have not been described in this way, and specific information regarding architectural design, branching order or configuration, vessel size, and characteristics of blood flow are lacking.

A. Rat Cremaster

The striated cremaster muscle, formed from external oblique and transverse abdominal muscles, is a thin, two-layered muscular pouch that encloses the testis and epididymus. When prepared for *in vivo* microscopic observation by dissection and spread on a suitable mount, the diameter is approximately 2.7 cm in rats that weigh between 80 and 110 gm (Fig. 3.1).

The thickness of the muscle layers varies in different regions, the medial and internal portions being somewhat thicker than the external portion of the spread pouch. In the thin regions, the muscle is between 150 and 180 μm, whereas the thicker regions measure between 166 and 200 μm. In the intact muscle pouch,

22　　　　　**3. Microvasculature of Specific Organs and Tissues**

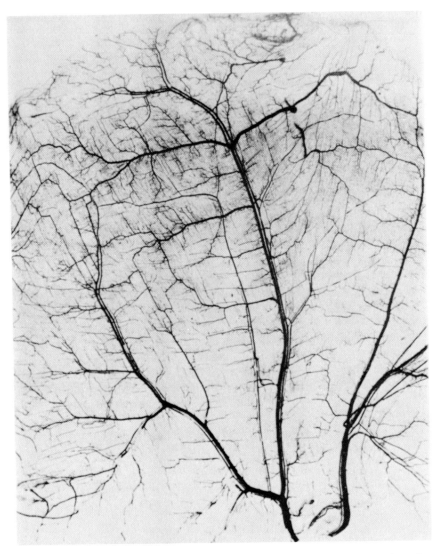

Fig. 3.1. Micrograph of cremaster muscle after injection of carbon intravenously to show vascular pattern. (From Smaje et al., 1970.)

the thin region corresponds to the anterior external area. No lymphatic vessels have been described.

The arterial blood supply to the muscle comes from branches of the external spermatic artery. A single paired artery and vein enter the pouch at the dorsal surface between the two layers of muscle which are oriented at 90° to one another. The diameter of the artery ranges between 110 and 130 μm, and its

1. Muscle

accompanying vein may be 150–190 μm. The artery occasionally divides into two and sometimes three vessels of smaller diameter (100–110 μm) at its entrance to the muscle, and these major arterial vessels then run longitudinally to the end of the muscle with a vein immediately adjacent. Branches from the major arteries, about eight to ten from each, have a diameter of 75–90 μm. Branches from these arterial vessels have been assigned diameters of about 40 μm and designated as second order vessels; the next order of branching (third) have diameters of about 10 μm. The fourth division of vessels, presumably arterioles, varies around 20 μm in diameter. Some of these arterioles give off arterioles that supply capillaries to both muscle layers (Fig. 3.2). Other arterioles of about 10μm, which may be called terminal arterioles in other preparations, divide into numerous capillaries. The capillaries, which are 5–6 μm at the arteriolar end and 6–7 μm at the venular end, run parallel to and also between the muscle fibers. The capillary vessels have numerous cross connections that occur not only between adjacent capillaries but also between capillaries in the two muscle layers. Three planes of capillary networks have been described—an internal, a middle,

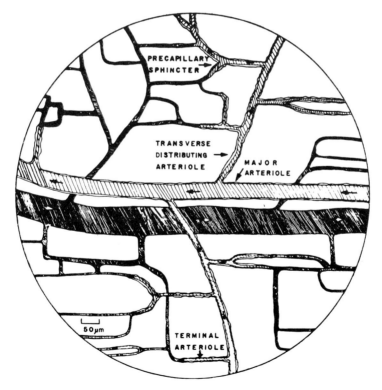

Fig. 3.2. Diagram of cremaster vasculature. Venous vessels are dark, arterial vessels are shaded. (From Hutchins *et at.*, 1974.)

3. Microvasculature of Specific Organs and Tissues

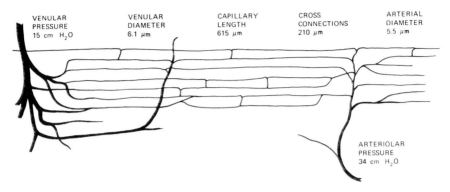

Fig. 3.3. Diagram of capillary bed in cremaster muscle and medium valves for measurements. Capillary density, 1300mm^{-2}; Distance between capillaries, 34 μm; capillary surface area, 244 cm^2/cm^3 muscle; micro-occlusion pressure data—arterial end, 32 cm H$_2$O; venular end, 22 cm H$_2$O; capillary filtration coefficient, 0.001 μm^3/μm^2 sec cm H$_2$O difference; and red cell velocity, 700μm/sec. (From Smaje et al., 1970.)

and an external layer, and the layers have been shown to interconnect. The basic capillary pattern can be described as a three dimensional network with vessels running parallel to muscle fibers and aligned in a ladderlike arrangement (Figs. 3.3 and 3.4).

Each artery has an adjacent vein through the first three or four branching orders, after which the vessels are separated and form a different vascular pattern. Both the arterial and venous vessels form arcades or anastomoses through-

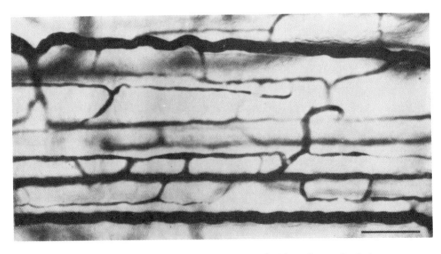

Fig. 3.4. Rat cremaster vessels injected with carbon and gelatin. Scale = 100μm. (From Majno et al., 1961).

I. Muscle

out the bed, although arteriovenous anastomoses are not seen. Only an occasional thoroughfare channel is seen according to two groups of investigators who use this preparation, although one author uses the term interchangeably with metarteriole. The arterial vessels that precede the capillary network do not have a pattern that conforms to the direction of the muscle fibers, whereas capillaries follow the direction of the fibers.

Blood flow through the capillary bed has been described as brisk, with intermittent flow produced by vasomotion of arterioles and precapillary sphincters. (Baez, 1973).

The nerves to the cremaster muscle enter at the neck of the sac, branching and intercommunicating as they go toward the tip. The greatest number of nerve fibers are myelinated ones that go to muscle end plates, while comparatively few unmyelinated fibers go to blood vessels. The genitofemoral nerve is the origin of most of the myelinated fibers that supply the two muscle layers. Strands from the ilioinguinal and iliohypogastric nerves also supply fibers. In addition, the lateral cutaneous nerve has been shown to contribute to the area. Surgical section of these four nerves will denervate the cremaster muscle and its blood vessels. Dilation of arterial vessels, exclusive of the minute vessels, follows surgical denervation but it subsides in a few days (see Fig. 3.5 for distribution of nerves).

B. Tenuissimus Muscle

The tenuissimus muscle in cats and rabbits is located in the dorsal part of the thigh and runs from the sacrum to the fascia of the lower leg. The muscle, which in the past has been used by neurophysiologists for afferent nerve studies (Bessou and Laporte, 1965), ranges in thickness from 0.3 to 0.6 mm in the central part of the muscle and 0.05–0.1 mm at the borders. The width of the muscle is 3–5 mm and its length approximately 10–12 cm.

The total number of muscle fibers, both red and white, average 1375 with an average diameter of 44.0 μm. The number of capillaries surrounding each fiber is approximately 3.6.

The muscle is supplied by an artery which does not supply any other muscles nor does its distribution anastomose with vessels from other surrounding muscles. The major artery entering the muscle has an average diameter of 110 μm and connects to the central artery which has a diameter on the average of 72 μm depending on the size of the animal and the muscle. The major draining vein (in some instances there are two) connects to the central vein. The central vein and artery run parallel to the muscle fibers about 25 mm apart and are accompanied by a nerve which has a diameter of 50–100 μm. First order branches from the central artery are perpendicular to the muscle fibers and have a wide range in diameter of 70–20 μm. These vessels are called transverse arterioles by Erikkson and Myrhage (1972), who assign them an average diameter of 22

Fig. 3.5. Nerve plexus on an arteriole in cremaster muscle. (From Grant, 1966).

I. Muscle

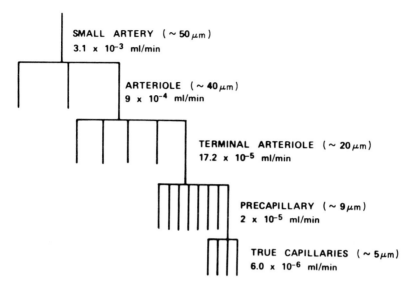

Fig. 3.6. Schematic of arteriolar branching derived from volume flow values. Ratio of flows in successive branches of small blood vessels provides an estimate of the number of branches at successive cross sections of the microvascular bed. (From Fronek and Zweifach, 1977).

μm. The transverse arterioles subdivide several times into a number of short precapillary vessels which then give rise to true capillaries. In most instances, each of the nine to ten precapillary vessels subdivide into three to five capillaries. Based on data of flow distribution, Fronek and Zweifach (1977) have proposed a schematic of arteriolar branching seen in Fig. 3.6.

The distribution of capillaries in the rabbit tenuissimus muscle is shown in Fig. 3.7.

Precapillary vasomotion is consistently observed and causes variations in flow velocity and vessel diameter by about 8–10%. Pressure measurements were also made by Fronek and Zweifach (1975), and it was reported that pressure in small arteries (50 μm diameter) was 96 mmHg; in arterioles (40 μm diameter) 88 mmHg; in terminal arterioles (20 μm diameter) 70 mmHg; and in precapillary vessels (9 μm diameter) the pressure had fallen to 38 mm Hg. Only about 30–35% of the capillaries had blood flow under resting conditions as determined by direct *in vivo* observation of the muscle, which is estimated to have a capillary density of 1000 vessels/mm^2.

Erikkson and Myrhage (1972) did not see any distinct precapillary sphincters, but describe 12 μm arterioles as having a single discontinuous layer of smooth muscle cells. More distally located vessels with a diameter of less than 10 μm had no muscle cells.

28   3. Microvasculature of Specific Organs and Tissues

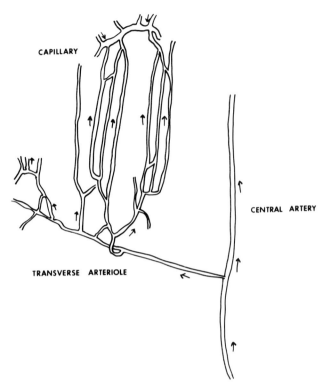

Fig. 3.7. A diagram of the distribution of capillaries from the transverse arteriole. (Courtesy of R. F. Tuma.)

Other investigators (Tuma *et al.*, 1975) did not identify precapillary sphincters at capillary origins and stated that sphincters (assuming that to be their only location) are not physiologically functional in the tenuissimus muscle. Vasomotion was described in small arterioles (20–40 μm) that affected groups of capillaries. The region of local control of blood flow into the capillaries of this bed remains to be accurately defined.

There is a lack of agreement about the presence of arteriovenous anastomoses in this skeletal muscle. Hammersen (1970) presents convincing arguments against their presence and suggests a new term ''arc-capillary'' (Bugelkapillare) to designate small vessels that are looplike and that represent a direct pathway between terminal arterioles and the smallest venules. The typical pattern as seen also in the rectus abdominus is shown in Fig. 3.8.

Erikkson and Myrhage (1972) note that arteriovenous anastomoses are rare exceptions, whereas Tuma believes them to be numerous at the periphery of the muscle. The belief is based on the observation that there are vessels between

I. Muscle

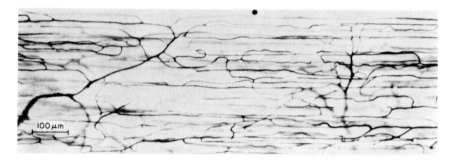

Fig. 3.8. Characteristic vascular pattern in M. rectus abdominis of the dog. (From Hammersen, 1968.)

arterioles and venules that are larger in diameter than most capillaries and have a faster flow. These vessels may be the same as the arc-capillaries described by Hammersen (1970).

A number of other skeletal muscle beds have been examined and described in more or less detail than the cremaster and the tenuissimus. For the most part, the vascular pattern of other muscles such as the spinotrapezius of the rat, the extensor hallicus proprius in the same animal, the gastrocnemius and plantaris in the monkey, the rectus abdominis in the dog, the abductor magnus of rabbit, and the gracilis muscle of the rat are very similar; therefore, detailed description of each seems unwarranted.

The rat extensor hallucis proprius as described by Myrhage and Hudlicka (1976) has a central artery and vein, transverse arterioles, and capillaries that run parallel to the muscle fibers and are connected by intercapillary branches. The vascular pattern is similar to that of the tenuissimus muscle. Terminal arterioles usually give rise to three or four capillaries, and capillary density was calculated to be 1247 capillaries/mm^2 at the proximal end of the muscle, with fewer at the distal end.

Although no vascular constriction was seen close to branch sites of capillaries, intermittent flow of blood in capillaries was observed, suggesting contractile activity upstream. The muscle is made up of mostly white fibers and Myrhage and Hudlicka believe it typifies muscles that have locomotion as their primary function.

The spinotrapezius muscle of the rat repeats the usual vascular pattern. The arterial and venous vessels show spontaneous vasomotor changes, and precapillary sphincters may close completely to curtail capillary blood flow when redistribution of blood is necessary (Zweifach and Metz, 1955). Gray (1971) gives average control diameters for six categories of vessels, ranging from a small artery of 55 μm to a vein of 113.7 μm.

The sartorius muscle of the cat follows the description in general, and observers report capillary flow, based on velocity measurements, varied between steady, periodic, and irregular (Burton and Johnson, 1972). Once again, the site of regulation of capillary flow was not identified. Henrich and Hecke (1978) have studied rat gracilis muscle and made continuous measurements of arteriolar diameters, red cell velocity, and intravascular pressures in second, third, and fourth order vessels. The capillary branching pattern and capillary lengths and diameters in the frog sartorius have been reported recently (Plyley *et al.*, 1976).

C. Myocardium

Descriptions of microvessels in heart muscle have been reported by two groups of investigators using *in vivo* preparations. Martini and Honig (1969) developed a method using stop-motion cinemicrophotography to study capillary blood flow in the beating rat heart under various conditions. Tillmanns *et al.* (1974) transmitted light through a 20 gauge needle inserted underneath the superficial layer of heart muscle of dogs. A system was designed to maintain the focal distance between the beating heart and the stationary microscope objective. Both investigations were directed primarily toward measurements of red blood cell velocity and changes in capillary diameter in the phases of systole and diastole. Descriptrons of microvessels in the myocardium derived from fixed material have appeared recently, notably by Bassingthwaighte *et al.* (1974), Grayson *et al.* (1974), and Phillips *et al.* (1979), but the vascular pattern formed by the microvessels between a distributing arteriole and a collecting venule in cardiac muscle has not been described. Certain aspects of the coronary vascular distribution have been observed using the various methods, and selected portions are repeated here.

Venous vessels appear most prominently on the outer surface of the heart in association with densely-packed capillary networks. The capillaries, which are quite long, run parallel to the myocardial muscle fibers and show short interconnections which are similar to those seen in skeletal muscle. The density of the capillaries is roughly 3500–4000/mm^2 (Fig. 3.9). Both *in vivo* and *in vitro* measurements of capillary diameter are in agreement that the mean diameter is around 4 μm in both rat and dog hearts (Sobin and Tremer, 1972). In the beating dog heart, capillary vessel diameters vary between 4.1 μm in systole and 6.3 μm in diastole (Henquell, 1976).

Arteriolar vessels of about 60 μm are seen to give rise to three arteriolar branches of 10 μm. The arteriolar vessels are relatively short and quickly divide into numerous capillaries. There are many more venous vessels formed by the confluence of capillaries than there are arteriolar vessels which precede the capillary micromesh. No arteriovenous anastomoses were seen in a fixed preparation.

I. Muscle   31

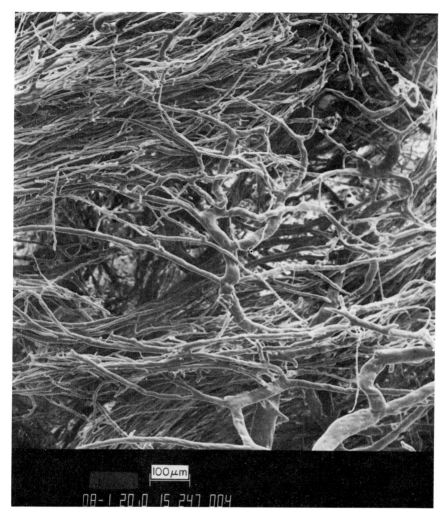

Fig. 3.9. Left Ventricular wall of cat heart injected with Mercox. Bar = 100 μm. (From Phillips *et al.*, 1979.)

In the living heart, areas designated as precapillary sphincters were seen to contract and relax in 3–5 minute cycles independent of arteriolar contractions. It has been suggested that the number of open capillaries was dependent on tissue P_{O_2} rather than on intraluminal pressure changes. Fixed preparations have shown a constricted area at the origin of an arteriolar branch that is referred to as a "precapillary sphincter-like structure" (Fig. 3.10).

3. Microvasculature of Specific Organs and Tissues

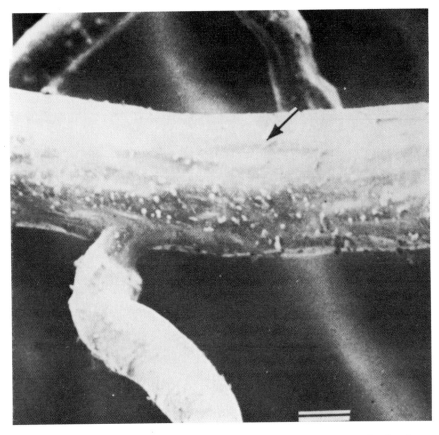

Fig. 3.10. Corrosion Mercox preparation of an arteriole in the left ventricle of cat heart. Bar = 100 μm. (From Phillips *et al.*, 1979.)

Although information regarding the microvasculature of the myocardium is still incomplete, it is remarkable, in view of the extremely high density of capillaries, the thickness of the muscle, and the variations imposed by the rhythmical contraction and relaxation of the heart, that a significant amount of descriptive material has accumulated Combining the knowledge gained from studies of *in vivo* and *in vitro* preparations, it seems permissible to suggest that cardiac muscle shares numerous features in common with skeletal muscle. It is extremely important, however, that the investigators of the coronary circulation use the current terminology common to other microvascular beds to avoid disagreements that are mainly semantic in nature.

II. VISCERAL ORGANS

A. Stomach

A complete description of the vascular pattern of the stomach from the muscle layer to the mucosa awaited the development of an *in vivo* microscopic technique. A means for microscopic observation of blood vessels in the gastric muscle, the submucosa, the muscularis mucosa, and the mucosa of the rat stomach was presented by Rosenberg and Guth (1970). The results of the microscopic studies using this technique follow.

The rat stomach is supplied with blood primarily from the left gastric artery and the right gastroepiploic artery. The left gastric artery divides into three branches, one of which goes to the anterior surface of the stomach, one to the posterior surface, and one to the lesser curvature. The anterior branch gives rise to branches that go to the forestomach, the proximal corpus, the distal corpus, and the antrum. The branch of the left gastric artery that goes to the lesser curvature of the stomach gives rise to four or five parallel branches that are 0.8–1.8 mm apart (Fig. 3.11). These branches pierce the muscle coat of the stomach and supply branches to the muscle layer. The right gastroepiploic artery, which goes to the greater curvature, gives off as many as ten to eleven parallel branches which are 1.2–1.8 mm apart. These arterial branches penetrate the muscle coat and continue to the muscle layer.

The arterial vessels that reach the muscle layer are 20–30 μm in diameter, and they divide into arteriolar branches which run in the muscle layer in a direction transverse to the long axis of the stomach. These arteriolar branches give off capillaries which run parallel to one another at a distance of 40–80 μm apart.

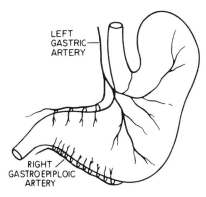

Fig. 3.11. Diagram of the major arterial blood supply to the rat stomach. The left gastric artery divides into three branches, which pass to the anterior, posterior, and lesser curvature surfaces of the stomach. (From Guth and Rosenberg, 1972.)

The capillaries are connected with each other on the same plane and with capillaries on different planes. The capillary network drains into venules which accompany the arterioles. The venules empty into veins which come from the submucosal plexus (Fig. 3.12).

The arterial branches that enter the muscle coat and give off branches to the muscle layer as just described continue and subdivide in the outer layer of the submucosa. These vessels form arterial arcades; the diameters of the vessels, now larger than the parent vessels, are 34–58 μm. From the arterial arcades, branches 20–40 μm in diameter arise and interconnect with each other to form a secondary arcade. These arcading vessels give off branches 14–19 μm in diameter, which are called mucosal arteries. The mucosal arteries pass through the muscularis mucosa and divide into two or three branches to form the mucosal capillary plexus.

Each of the three or four branches of the mucosal arteries give rise to three to six capillaries at the base of the mucosa. Mucosal arteries are connected by thin anastomosing channels. The capillaries of the superficial mucosa have a hon-

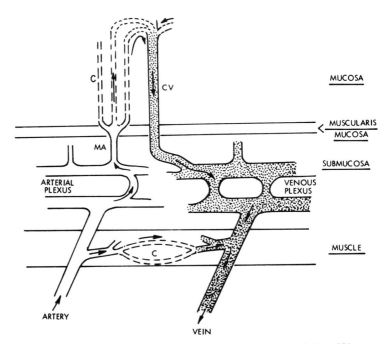

Fig. 3.12. Diagrammatic representation of the gastric microcirculation. MA, mucosal arteriole; C, capillary; CV, collecting vein. The microvasculature of the muscle layer is in parallel with that of the mucosa, while the microvasculature of the submucosa is in series with that of the mucosa. There are no arteriovenous anastomoses. (From Guth, 1977.)

II. Visceral Organs

eycomb appearance because of their distribution around the openings of the gastric glands (Fig. 3.13).

Collecting veins are distributed along the mucosa and the blood can be seen flowing into them from the honeycomb of capillaries.

The venous pattern of the submucosa is similar to the arterial pattern. Veins from the mucosa run perpendicularly into the submucosa and therefore appear as round dots when viewed through the submucosa. The diameter of these venous vessels is 30–50 μm. They empty into veins in the submucosa to form the submucosal venous plexus, and they are interconnected in the same way as the arterial vessels that form the arterial plexus. As the veins leave the submucosa, branches from veins of the muscle layer join them.

Contrary to older descriptions obtained from fixed tissue or dye injection studies, no arteriovenous anastomoses or a-v shunts were found in the gastric circulation. More current studies involving the use of radioactive microspheres support the idea that no significant degree of a-v shunting occurs in this bed.

The blood flow through the main arteries which supply the stomach is continuous and in one direction. Change in the direction of flow is observed in the arterial arcades of the submucosa when flow from one parent artery dominates the flow from a connecting parent artery.

B. Intestine

The superior mesenteric artery is the main source of blood supply to the small intestine. The blood supply to the various layers of the small intestine is brought to the outermost layer of intestinal muscle through small arteries of the mesentery. The descriptive material of the vasculature within the intestinal wall which follows is based on *in vivo* microscopic observations by Baez (1977), Gore and Bohlen (1975), Bohlen *et al*. (1975), and Bohlen and Gore (1976, 1977).

A small artery in the mesentery that runs parallel to the intestine gives off an arteriolar branch that penetrates both the outer longitudinal muscle layer and the inner circular muscle layer and enters the upper surface of the submucosa. These arterioles, which are 60–80 μm in diameter, then branch to form second order arterioles which run parallel to the longitudinal axis of the intestine. There are three to five second order arterioles from each first order arteriole. From each second order arteriole, 16 to 18 branches are given off which, as third order arterioles, go through the submucosa toward the innermost layer and form the central arterioles of the mucosal villi. The third order arterioles also give off branches which carry blood out toward the muscular layers to the plane of cleavage between the two muscle coats. Here, the fourth order arteriole gives off smaller fifth order arterioles which run perpendicular to the muscle fibers and serve as parent vessels (terminal arterioles) for two sets of capillaries. One group

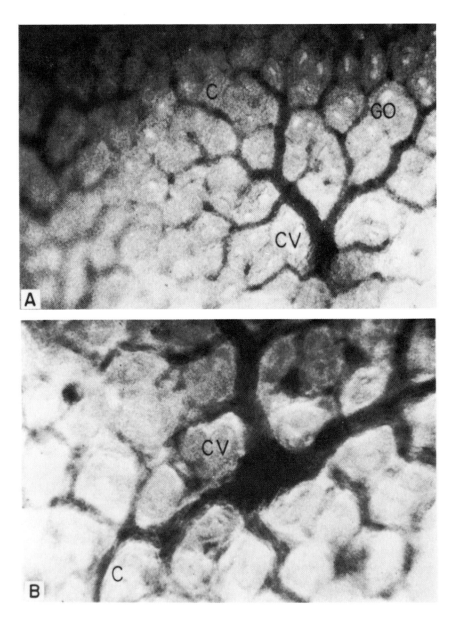

Fig. 3.13. (A) Photomicrograph of the rat mucosal preparation. The honeycomb-like appearance of the capillaries (C) surrounding the glands and ultimately draining into collecting veins (CV) is clearly observed. The orifices (GO) of the glands in the center of surrounding capillaries can be seen in some areas. (Original magnification ×100). (B) Photomicrograph of the rat mucosal preparation at a higher magnification. The features described in (A) are more readily seen. (Original magnification ×200). (From Guth, 1977.)

II. Visceral Organs

of capillaries runs in the same plane and supplies the circular muscle, and the other group continues out to the longitudinal muscle layer. This second set of capillaries runs between the muscularis and the serosa, supplying them and the longitudinal muscle layer with blood (Fig. 3.14).

Some of these smallest arterioles have a smooth muscle investment designated as a precapillary sphincter. The capillaries communicate with one another in their parallel course through the muscle fibers by lateral branches and with capillary vessels in an adjacent layer of muscle. The capillary vessels in the intestinal muscle layer are quite long, being between 700 and 950 μm in length. They are perhaps the longest of the pure endothelial (capillary) vessels in any vascular bed.

The mucosal arteries, which are interconnected with one another through anastomosing branches, have one or two arteriolar branches leaving at right angles before the mucosal artery enters the villus. These lateral branches then break up into six or eight endothelial capillaries. The terminal part of the mucosal artery, which has a smaller diameter than at its origin, enters the villus either as a single vessel or branches into two vessels.

The venous outflow paths begin where capillaries of the muscle layers join fourth order venules which converge to form third order venules. The third order venules go through both muscle layers toward the submucosa where they join

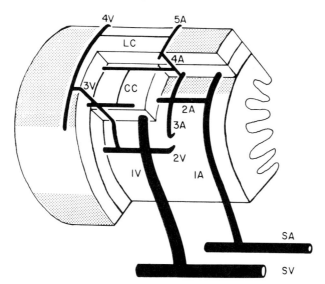

Fig. 3.14. Diagram of rat intestinal muscle microcirculation showing a single vascular unit. Vessels classified numerically according to branching. Five major arteriolar branches (1A–5A). Four major venular branches (4V–lV). Capillaries (CC) in inner circular muscle and capillaries (LC) in outer longitudinal muscle layers run parallel to muscle fibers. (From Gore and Bohlen, 1975.)

second order venules, vessels which are receiving blood from the collecting venules of the mucosal villi. The second order venules drain into first order venules, which leave the small intestine to empty into the small vein in the mesentery. This vein accompanies the mesenteric artery from which the blood supply to this vascular unit originated. Both the arteriolar and venular vessels in this preparation exhibit cyclic constriction and relaxation, and this vasomotion is considered to represent a normal and characteristic behavior of these vessels. The pressures and diameters of the blood vessels described are shown in Table 3.1.

The distribution of vessels within a villus begins with the main arteriole that enters the base and continues to the apex. Here it divides into two smaller vessels designated as distributing arterioles. These descend into the villus, giving off no branches for a short distance after the division. Capillary offshoots then appear, each encircled by a single vascular smooth muscle cell at their origin, and they form a network under the mucosal epithelium and in the interior portion of the villus. Five to ten capillaries join and form second order venules on either side of the interior of the villus. The two collecting venules join to form a large venule at the base. A diagrammatic representation of this arrangement is shown in Fig. 3.15.

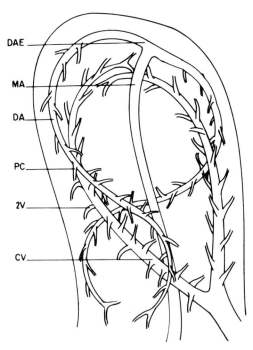

Fig. 3.15. Vascular anatomy of a villus. This diagram demonstrates spatial relationship of microvessels within a villus. MA, main arteriole; DAE, distributing arteriole entrance; DA, distributing arteriole; PC, precapillary sphincter; 2V, second order venule; and CV, collecting venule. (From Bohlen *et al.*, 1975.)

TABLE 3.1

Micropressure and Diameter Distribution in Rat Intestinal Muscle and Mucosal Microcirculation[a,b]

Vessel category	Pressure (mm Hg)	Inside diameter (μm)	N
Muscle circuit			
First-order arteriole	44.6 ± 1.6	52.6 ± 1.8	16
Second-order arteriole	44.6 ± 2.9	29.6 ± 2.3	11
Third-order arteriole	32.4 ± 2.5	12.2 ± 0.6	14
Fifth-order arteriole	26.7 ± 2.0	8.4 ± 0.5	11
Capillary	23.8 ± 1.5	5.0 ± 0.5	8
Fourth-order venule	15.2 ± 1.2	9.9 ± 0.8	20
Second-order venule	15.7 ± 1.2	28.3 ± 5.4	10
First-order venule	10.1 ± 0.4	60.4 ± 3.6	11
Mucosal circuit			
Distributing arteriole	30.5 ± 1.7	8.3 ± 0.3	14
Capillary	13.8 ± 2.2	4.3 ± 1.1	11
Second-order mucosal venule	12.8 ± 1.5	9.3 ± 0.6	8

[a] From Bohlen and Gore (1976).
[b] Systemic arterial pressure was 100–110 mm Hg. Data are expressed as mean ± SE.

C. Mesentery

1. Cat

The mesentery, a thin sheet of tissue which suspends the intestines and helps hold them in place, is traversed by large arterial vessels that supply blood to the intestinal wall. Some of the blood in these vessels, which originates from the superior mesenteric artery, is diverted to supply blood to the mesenteric membrane and to the adipose tissue commonly distributed throughout the mesentery. The venous vessels which carry blood from the tissue eventually empty into the portal vein.

In considering the vascular pattern displayed on this thin, voluminous tissue, it is convenient to divide the mesentery into portions delineated by three distinctly different sizes of arterial and venous vessels. In viewing a portion of the mesentery, the largest or first order vessels form a triangular pattern. The paired arterial and venous vessels extend from their posterior attachment toward the intestine where they bifurcate and then run parallel to the intestine. These large vessels, which are 150–300 μm in diameter, outline a sector (Fig. 3.16).

A second pattern is formed by the paired arterial and venous vessels within each sector. The second order vessels are considerably smaller, being between 20 and 40 μm, and they form a network arranged in such a way that small circumscribed areas of varying size and shape are seen. These smaller areas within each sector are referred to as modules (Fig. 3.17).

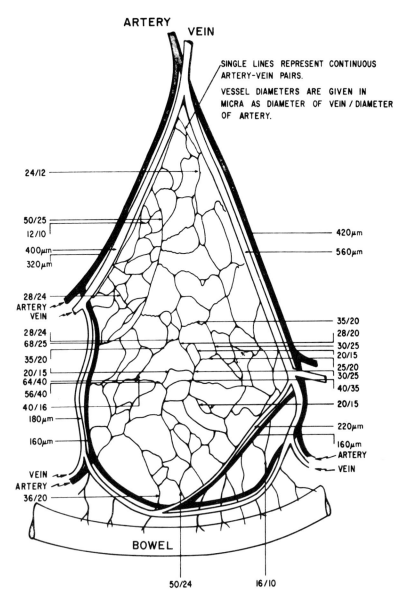

Fig. 3.16. Schematic representation of a typical sector of cat mesentery. (From Lipowsky and Zweifach, 1974.)

II. Visceral Organs

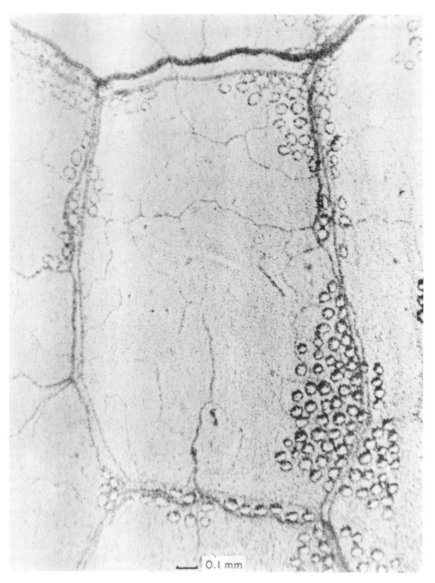

Fig. 3.17. Single frame photograph of a module. (From Frasher and Wayland, 1972.)

3. Microvasculature of Specific Organs and Tissues

A module is defined as an area of membrane completely surrounded by pairs of arterial and venous vessels that form artery to artery and vein to vein arcades at the apices. The interior of the module is filled with a dense network of precapillary, capillary, and postcapillary vessels. Zweifach (1974a) has described these vessels in the terminal vascular bed of the cat mesentery as follows. Small arteries have diameters of approximately 50 μm and walls that are 2-3 μm thick. Arterioles, the next order of branches, are vessels with diameters of 20-40 μm, with a single layer of smooth muscle. The final length of 200-300 μm of the arterioles is designated as terminal arteriole, and precapillary vessels are the lateral branches from the arterioles from which capillaries originate. The venous system begins where capillaries join to form the first of the confluent vessels. These venous vessels have diameters of 15-20 μm and are called postcapillaries. The postcapillaries converge to form vessels which are still nonmuscular and are called collecting venules. Collecting venules in turn form muscular venules, 40-60 μm in diameter, which are accompanied by an arteriole. The capillary bed is considered to be the entire array of vessels between a terminal arteriole and a collecting venule (Fig. 3.18).

Extensive studies of microcirculatory structure and function of the cat mesentery by Zweifach (1974a,b; Zweifach and Lipowsky, 1977) have resulted in information about this particular bed other than the vascular pattern. Variation in

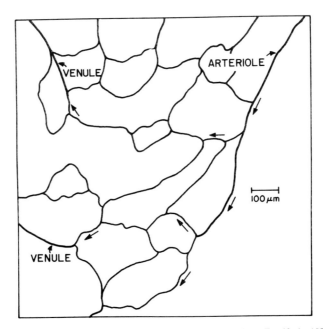

Fig. 3.18. Capillary network in cat mesentery. (Redrawn from Zweifach, 1974b.)

II. Visceral Organs

pressure occurs in many vessels as the result of spontaneous changes in vessel diameters. A parent arteriolar vessel and a branch from it can show changes in diameter independently of one another. Precapillary branches constrict or dilate resulting in a 20–25% change in diameter. Vessels extending from the arterial to the venous side that act as anatomical shunts (referred to as thoroughfare channels in some beds) are less common in the cat mesentery than in mesenteries of rat, rabbit, and dog. Shunt vessels seen in the cat mesentery are characterized by a more rapid flow than interconnecting vessels and are wider than other capillaries in the bed.

The cat mesentery is richly supplied with lymphatic vessels. Their structure and characteristic behavior differs only minimally from lymphatics in other mesenteric beds. The smallest lymphatics are thin-walled endothelial vessels that are much larger than the capillary vessels in the same tissue, having diameters between 25 and 50 μm. The terminal lymphatic vessels are actually wider than the collecting venules, even though they are frequently referred to as lymphatic capillaries. Because this terminology seems inappropriate, it has been suggested that "terminal lymphatics" be used to refer to the lymphatic network that is involved in the exchange of materials in the interstitial spaces, while "collecting lymphatic channels" be used to designate the large vessels of 200 μm diameter that are formed by confluences. These collecting vessels have valves along their course to assure flow in one direction.

The majority of terminal lymphatics in the mesentery form an interconnecting network that follows the arcuate pattern of the blood vessels to some extent. They follow the paired arterioles and venules rather than the capillaries. The walls are extremely thin, consisting of a delicate endothelial membrane. There are a number of blind lymphatic terminals which show no distinctive pattern. Their bulbous ends may lie free in the interstitium or may lie against a venule. A thin endothelial barrier always separates the terminal end of the lymphatic vessel from the blood vessel.

As the networks of terminal lymphatics converge, they form collecting channels which have diameters between 40 and 50 μm and which increase in size to diameters of 200 μm. The wall of the collecting channels have a thickness of several microns and the channels have valves at intervals of several hundred microns. Valves always appear where two collecting channels converge. Although the valve leaflets appear to be very thin and therefore fragile, they exhibit considerable resistance to retrograde pressure. Figure 3.19 depicts the terminal lymphatic network with a collecting channel.

In contrast to lymphatic vessels in rat and guinea pig mesentery, the lymphatic vessels in cat mesentery show no spontaneous contractile activity, nor is it possible to evoke contractions by mechanical or chemical stimulation. This lack of contractile activity in lymphatic vessels of the cat mesentery would seem to be unusual for lymphatic vessels and suggests the lack of an investment of any contractile tissue on these particular lymphatic vessels.

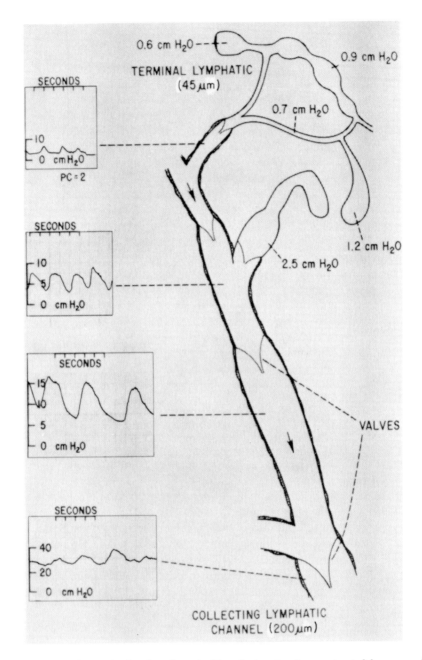

Fig. 3.19. Composite drawing of terminal lymphatic network reconstructed from several mesentery preparations. (From Zweifach and Prather, 1975.)

II. Visceral Organs

2. Rat

The cecal mesentery of the rat was one of the first of the mammalian structures to be used for microscopic observation of small blood vessels *in vivo*, having been introduced by Chambers and Zweifach (1944). It has been widely used for studies of anatomical, physiological, and pharmacological features of the microcirculation.

The cecal mesentery is a very thin tissue that holds the blind end of the cecum to the adjacent intestine and receives its vascular supply from blood vessels that also supply the intestinal wall (Fig. 3.20).

The larger vessels are paired, and the arterial vessels are about one-third the size of the accompanying vein. Arterial and arteriolar blood flow is rapid, venous flow is steady and continuous, and capillary flow is intermittent with a definite rhythmicity. The intermittent capillary flow can be seen to result from contraction of precapillary sphincters. Precapillary sphincters are seen at all branchings of capillaries from muscular arterioles. The sphincters are identified as the muscular proximal portion of the arteriolar offshoots which then continue as the true capillary network of endothelial tubes with no perivascular muscle. A prominent structural feature of this vascular bed is the thoroughfare or preferential channel, which consists of a direct interconnection between the arterial and venous vessels. The preferential channel has been identified in some tissues besides the

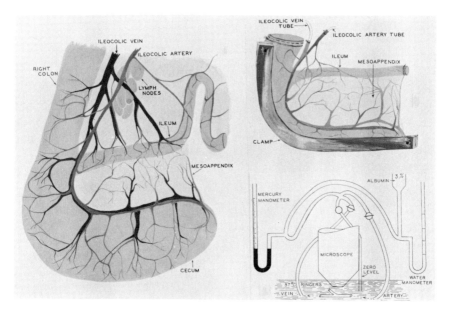

Fig. 3.20. Diagram of rat cecal mesentery showing origin of vascular supply. (From Baez, 1961.)

mesentery, although they are much less numerous. It is totally absent in other tissues. The venous end of the preferential channel receives blood from inflowing capillaries. A venous collecting system is formed by the confluence of capillaries which originated from the muscular arterioles. Some investigators refer incorrectly to the precapillary sphincter, described by Zweifach as the muscular portion of a capillary offshoot, as a vessel rather than a specific area of a vessel. A typical vascular pattern as seen in the cecal mesentery is found in Fig. 3.21. The vessel designated as a preferential channel appears as a loop formed by the arteriole with distributing branches. The vessel continues and becomes a venule with inflowing vessels. This pattern could be characteristic of a thin, relatively avascular tissue such as the cecal mesentery, where the major supplying and draining vessels are found at the periphery of the tissue and give off vessels that invade the mesentery and then return to the periphery.

The arterial components of this vascular bed show a distinct gradation of

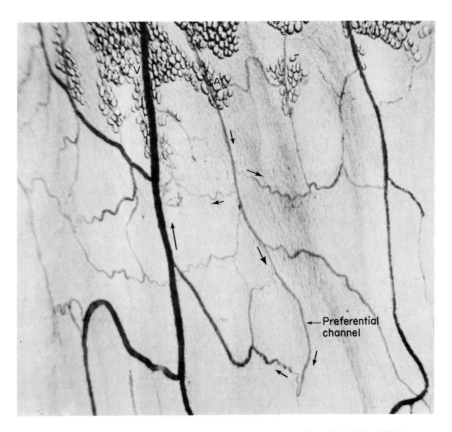

Fig. 3.21. Pattern of vessels in rat cecal mesentary. (From Zweifach, 1954.)

II. Visceral Organs

reactivity to applied vasoactive substances, with the greatest responsiveness seen in the precapillary sphincter. There is a progressive decrease in responsiveness back toward the arterioles. Vasomotion, the spontaneous periodic narrowing and dilation of vessels, is seen in the muscular arterioles and their branches.

D. Liver

The highly specialized and numerous functions of the liver necessitate a complex vascular system in order to deliver blood from two sources to be utilized by liver cells. The portal vein transports food material from the intestine to be processed in the liver, while the hepatic artery, formed after a series of branches from the aorta, transports arterial blood to maintain cellular life. There are, then, two separate afferent systems that carry blood into the liver lobule.

The middle of each hepatic lobule is occupied by a central venule. The central venule receives blood from the sinusoids, which are oriented at right angles to it. The horizontally aligned sinusoids are connected to each other by vertical intersinusoidal sinusoids. The portal venules run parallel to the central venule at the periphery of the lobule and are connected to the central venule through the sinusoids. These vessels, which constitute the drainage pathways of each lobule, are shown diagrammatically in Fig. 3.22.

Arterial blood is distributed through the hepatic arteriole, which parallels the portal vein close to the bile ducts and lymphatic vessels, and gives off numerous branches in its course through the lobule. The terminal portion of the arterial

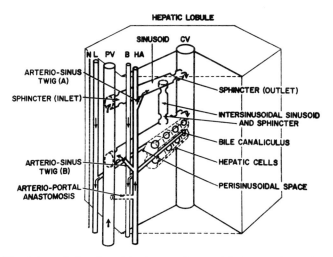

Fig. 3.22. Schematic drawing of the functional unit of the liver and its relationship with the hepatic lobule. PV, portal vein; HA, hepatic arteriole; N, nerves; L, lymphatic; B, bile duct; CV, central venule. (From McCuskey, 1966.)

system of the liver is made up of four different types of vessels according to Burkel (1970). The first of these are the arterioles, which are usually five times smaller than the portal vein that they accompany. Diameter measurements of hepatic arterioles have been obtained from a guinea pig *in vivo* preparation by Irwin and MacDonald (1953), from a rat *in vivo* preparation by McCuskey (1966), and from fixed tissue in the rat by Burkel (1970). The diameters are widely variable in size and to report the sizes would serve no useful purpose in this description. The structural makeup of the arterioles is not unusual. They have an endothelial lining, an elastic interna, two layers of encircling smooth muscle cells with numerous unmyelinated nerves in the outer layer, and a basement membrane.

Terminal arterioles branch from arterioles at acute angles and continue to become capillaries. The terminal arterioles are simpler in construction than arterioles in that they lack an elastic interna and have only one layer of smooth muscle cells that encircle them. The number of unmyelinated nerves is sometimes greater than those seen at the level of the arterioles. The terminal arterioles give off capillaries which supply the tissues of the portal space. Around the origin of the capillaries are smooth muscle cells which are referred to as "precapillary sphincters" even though they are innervated. The smooth muscle investment continues for some distance along the vessel wall until it becomes incomplete and finally disappears. The vessel then satisfies the usual definition of capillary.

The capillaries are thickly distributed throughout the portal space and terminate by entering venous vessels or sinusoids. *In vivo* studies by Rappaport (1958, 1973) and McCuskey (1966) have described the manner in which the terminal vessels enter the venous system and sinusoids (Fig. 3.23).

The vessels in the liver contract and relax independently of one another. This spontaneous activity has been seen in hepatic arterioles, portal venules, central venules, and sinusoids. In addition, it has been shown that the endothelial cells which are located at the junction of sinusoids and portal venules (inlet sphincters), and also the endothelial cells located at the junctions of sinusoids with portal venules (outlet sphincters), can enlarge and bulge into the lumen of the vessel to effectively reduce the internal diameter. If two endothelial cells are located opposite to one another, they can completely occlude the lumen (Fig. 3.24).

E. Spleen

For a period of 30 years after Knisely's (1936) paper describing the circulation of blood through the spleen of a living mammal, support could be found for the premise that blood in the red pulp flowed through endothelial-lined channels, and therefore the splenic circulation was a closed system. This was not

II. Visceral Organs

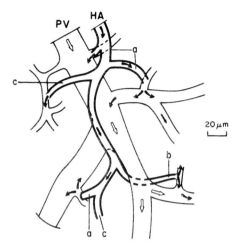

Fig. 3.23. Hepatic arteriole (HA) and an adjacent portal vein (PV). (From McCuskey, 1966.)

a universally accepted concept since there was evidence to suggest that the blood in the red pulp flowed through a meshwork formed by the processes of reticular cells and reentered the vascular system through large openings in the walls of the venous sinuses. Most recent studies by McCuskey and McCuskey (1977) support this latter view of an open circulation of blood through the red pulp of the spleen, which they describe as follows: A central arteriole leaving the white pulp be-

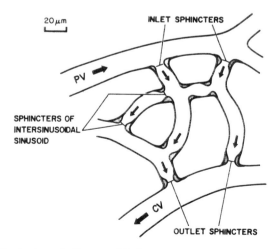

Fig. 3.24. Semi-schematic diagram of the location of hepatic sphincters, portal vein (PV), and central venule (CV). (From McCuskey, 1966.)

comes devoid of smooth muscle and ends in the red pulp as an arterial capillary. From these capillaries, blood flows into the marginal sinus or into the red pulp through a mesh formed by the processes of reticular cells (Fig. 3.25). The arrangement of the processes of the reticular cells in relation to the endothelium of the arterial capillaries may give the appearance of an endothelial-lined channel when observed through the microscope in the living animal. The arterial capillaries may connect directly with sinuses or venules but with an interrupted endothelial boundary.

Blood leaves the red pulp by re-entering sinuses and venules through large openings in the endothelium of these vessels. The openings are frequently large enough to permit red blood cells to pass through without any deformation.

Reilly and McCuskey (1977) reported dimensions for various vessels observed in the mouse spleen. The values are given here for the purpose of comparison. Central arterioles of the white pulp, which were identified by their walls with endothelium and smooth muscle, have an inside diameter of 6.4–11.2 μm. The flow of blood through these vessels is very rapid. Branches from the central arteriole, which leave the white pulp to enter marginal zones of the spleen, have an inside diameter of 4.8–8.0 μm. Terminal segments of these vessels become

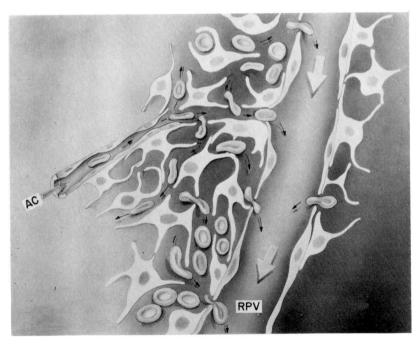

Fig. 3.25. Diagram illustrating the circulation of blood through the red pulp of the mouse spleen. AC, arterial capillary; RPV, red pulp vein. Arrows indicate the direction of flow. (From McCuskey and McCuskey, 1977.)

III. Lung

the arterial capillaries of the red pulp and have an inside diameter of 3.2–5.6 μm. The channels in the red pulp into which these arterial capillaries empty have inside diameters of 4.8–11.2 μm. The arterial vessels were shown to be innervated by adrenergic nerves by these investigators and the vessels showed the expected responses to various vasoactive drugs.

III. LUNG

The final ramification of the pulmonary vascular bed is difficult to study microscopically in the living animal because it is relatively inaccessible without extensive surgical preparation and also because of the continous movement with inspiration and expiration. Some ingenious methods have been devised, however, to permit visualization of small areas, and from these observations, characteristics of blood flow in the active lung have been described.

The architecture of the pulmonary microvasculature has been determined by studies on fixed tissue examined by light and electron microscopy, most recently by Rhodin (1978). His description from studies on cats is presented here. Small pulmonary arteries are muscular vessels that have numerous layers of smooth muscle cells, circular and spiral in distribution around the vessels that accompany the respiratory bronchioles. Nonmyelinated nerve fibers are found in the adventitia (Fig. 3.26).

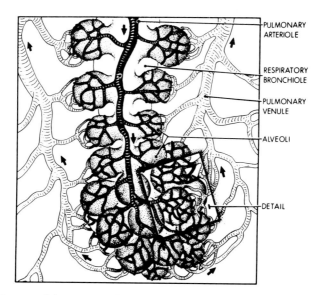

Fig. 3.26. Diagram of the microcirculatory bed of the respiratory bronchiole, alveolar sac, and individual pulmonary alveoli. (From Rhodin, 1978.)

Branches from the pulmonary arteries are called pulmonary arterioles and have a smaller diameter. They branch directly over the length of the small pulmonary arteries at a 90° angle and constitute the end of the pulmonary arterial system where the small pulmonary artery ends. As in the small pulmonary arteries, the arterioles possess several layers of smooth muscle cells, and the adventitia contains nerve fibers. Terminal arterioles are formed from the pulmonary arterioles and are called the precapillary sphincter area by Rhodin (Fig. 3.27).

The wall of these vessels has a thin endothelial lining and is covered by one or two layers of smooth muscle cells. Occasionally a nerve fiber is seen. The smooth muscle cells disappear once the terminal arteriole divides into a dense network of capillary vessels (Fig. 3.28).

The pulmonary capillaries of cats, which have diameters of 5–10 μm, form a dense network covering the alveoli lying within the interalveolar septa. One capillary may serve several alveoli. The covering of capillaries is so dense that the spaces between capillaries are often smaller than the capillary diameter. The observation of multinumbered vessels with so little space in between resulted in the concept of the interalveolar microvascular sheet (Fung and Sobin, 1969; Sobin *et al.*, 1970).

Postcapillary venules originate adjacent to the alveoli within the interalveolar septa. They are short, stubby vessels, slightly larger in diameter than the capillaries. Postcapillary venules have a sparse supply of smooth muscle cells which increase in number as the postcapillary venules converge to form pulmonary venules.

In vivo microscopic studies by Wearn *et al.* (1934) and by Irwin and co-

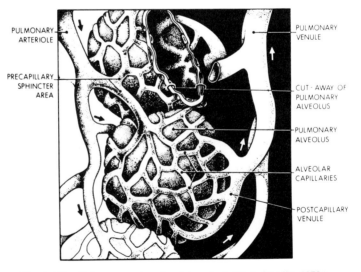

Fig. 3.27. Pulmonary microcirculatory bed. (From Rhodin, 1978.)

IV. The Bulbar Conjunctiva

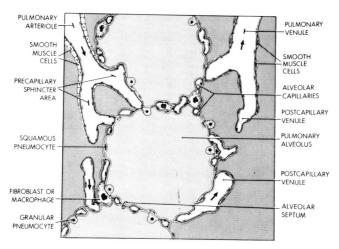

Fig. 3.28. Simplified drawing showing the relationship of arterioles, capillaries, and venules to the intra-alveolar septa. (From Rhodin, 1978.)

workers (1959) supplied information regarding the flow of blood in the pulmonary microvessels. Arteriolar vessels show a pulsatile flow at times and a steady flow at others. The type of flow may vary between two arteriolar vessels arising from the same arteriole. Contraction and relaxation of the arterioles with frequent reversal of flow is not uncommon. Arteriolar diameter changes are more marked than changes in diameters of venules. It is postulated that the intermittance of blood flow in the pulmonary microvessels depends on spontaneous contractile activity of the investing smooth muscle cells rather than as a response to neural excitation.

IV. THE BULBAR CONJUNCTIVA

The bulbar conjunctiva is a highly vascularized, thin, transparent mucous membrane on the anterior surface of the eye. It extends from the palpebral membrane, which lines the eyelids over the exposed surface of the eyeball, to the outer margin of the cornea. The conjunctiva is one of the few places where small blood vessels can be viewed under magnification in man. This type of examination has been used for the purpose of diagnosis for imminent disease states such as hypertension and diabetes in human patients (see Davis and Landau, 1966; Ditzel and Heuer, 1975). Pure descriptive material relative to the vascular pattern and activity of the vessels comes also from nonclinical studies.

Lee and Holze (1950), responding to the new and stimulating findings of Chambers and Zweifach (1944) in the microcirculation of the mesentery and omentum of rats and dogs, studied the arrangement of terminal arterioles, capillaries, and venules in the human conjunctiva and found the pattern to be very similar to that of animals. Using a standard type slit-lamp microscope with a magnification of 62×, they identified arterioles, collecting venules, true capillaries, precapillary sphincters (or regions) that were reactive, and main arteriovenular channels that formed a direct connection between arterial and venous vessels. Spontaneous vasomotor activity was seen to occur at the precapillary region. The vasomotion was characterized by complete constriction of the arteriolar vessels lasting for 2–3 minutes, followed by a 1–5 minute period of relaxation before the next contractile phase. The velocity of blood flow was found to average 0.12 mm/sec in arterioles which were 10–14 μm in diameter.

Grafflin and Corddry (1953) supplied additional information in 1953, emphasizing that an endless variety of vascular patterns could be found in the human conjunctiva. They could not identify a specific structural and functional unit, but did see numerous arteriovenous connections with such frequency that they believed them to be a characteristic feature of this bed. Connections between veins and artery-to-artery connections were also common (Figure 3.29). Bloch (1956)

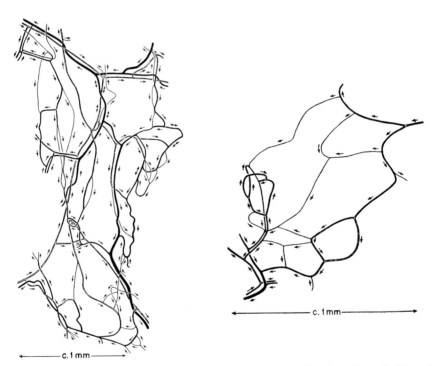

Fig. 3.29. Superficial vascular patterns seen in human conjunctiva. (From Grafflin and Corddry, 1953.)

published a detailed descriptive paper on circulating blood in the conjunctiva in health and disease. He points out that arteriolar vessels should be easily identified by noting the direction of flow, which is always toward progressively smaller vessels until the capillaries are reached. He noticed frequent changes in the direction of flow, a characteristic also seen in other vascular beds such as mesentery and omentum. Bloch found that the most superficial arterioles were usually deeper in the tissue than their corresponding venules, that the course of the arterioles is straighter than that of their accompanying venules, and that the arteriolar branches leave their parent vessels at a smaller angle than the venular vessels entering a confluence, which is usually at a right angle.

Fenton et al. (1979) observed conjunctivas in normal individuals for comparison with diabetics. They described the microvascular network of normal conjunctivas observed at 40× as having feeding arterioles with an inside diameter of 10 μm and very thin walls. These vessels are easy to recognize because they are long and straight and accompanied by a venule that is much wider in diameter. Two or three such arterioles usually supply the capillary network of an area 11 × 15 cm. This same capillary network is drained by four or five venules with diameters from 40 to 90 μm. All vessels in the conjunctival beds with diameters above 15 μm are venules. The slightly larger venules, those in the 25–40 μm range, may connect with one another but most frequently join to form larger vessels which may have diameters between 75 and 90 μm. The capillary bed is formed by randomly distributed second and third order branches from the feeding arterioles. The greatest number of vessels in the conjunctival vasculature are the capillaries and the postcapillary venules.

V. PIA MATER

In present day studies, the pial circulation in the living animal is viewed through the microscope most frequently in cats, rabbits, and rats, using either cranial windows or open craniotomies. A detailed description of the vascular pattern seen on the surface of the brain is lacking, but information is available from investigations concerned with blood flow, blood pressures, and vascular reactivity. Forbes (1928) noted that the larger arteries pulsated with cardiac systole, that the color contrast between the bright scarlet arterial vessels and the purplish red veins was striking, and that there were numerous small arterioles which either joined other arterioles or turned down into the cortex and disappeared. This abrupt turn gives the appearance of small circular dots on the surface of the brain. Reversal of flow in arterioles is frequently seen, but no contractile activity is associated with a change in flow, nor are spontaneous diameter changes in arterial vessels seen. Rosenblum and Zweifach (1963) reported that arteries of the mouse pial circulation are much smaller than their corresponding veins and seem to be more superficially oriented. Artery to artery and vein to vein connections were identified, but no arteriovenous anastomoses were seen. Perforating arteries and veins were studied in fixed sections that showed extensive capillary networks below the surface. These networks connected the terminal beds that originated from perforating arteries (Fig. 3.30).

Spontaneous flow changes are seen to occur in both arterial and venous vessels. Using *in vivo* microscopy Ma *et al.* describe rat pial vessels, and they found many more venules with tributaries than arterioles. They did not find the parallel distribution of arterioles and venules as seen in other beds. The diameters of six orders of vascular and arteriolar vessels as reported by them are presented here for comparison of relative sizes of arterial and venous vessels of the same branching order. The average diameter of the largest arterial vessel was 49.2 μm and the venous vessel was 100.3 μm. The average diameters of branches from this arterial vessel in sequence were 34.4 μm, 23.4 μm, and 6.8 μm and the arterial capillary was 4.2 μm. Returning on the venous side, the venous capillary had an average diameter of 4.0 μm, and the joining venous vessels were 7.9 μm, 14.3 μm, 32.7 μm, 56.9 μm, and finally 100.3 μm. The body weight of these rats was approximately 167 gm. Navari *et al.* (1978) reported no change in diameters of arterial vessels ranging between 30 and 300 μm in the exposed cortex preparation of cats over a period of 5 minutes, but noted that the resting diameters in the cranial window preparations were significantly larger than in the open preparation. The vascular pattern as seen in pial circulation of a rabbit is shown in Fig. 3.31.

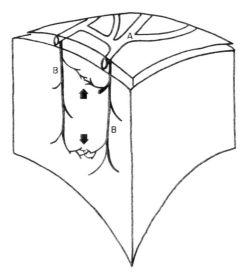

Fig. 3.30. A schematic wedge of the cerebral cortex showing anastomosing pial arteries (A) which give off branches (B) that penetrate the cortex. The branches are less than 60 μm in diameter and ramify to form the capillary bed indicated by arrows. (From Rosenblum, 1965.)

VI. Special Tissues

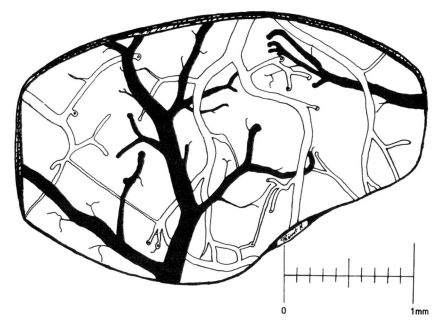

Fig. 3.31. Pial circulation of the rabbit. The penetrating arterial vessels and the emerging venous vessels appear as circles on the surface of the brain. (Courtesy of Dr. S. Reuse-Blom, University of Libre, Brussels, 1979.)

VI. SPECIAL TISSUES

A. Hamster Cheek Pouch

Cheek pouches of the hamster are internal, bilateral sacks that lie beneath the skin and extend from a slit-like opening in the corner of the mouth down to the shoulder region, a distance of about 4 cm. The empty and relaxed pouch is approximately 3–5 cm long and 1 cm wide. When everted and spread over a mount suitable for microscopic observation, the dimensions can be extended to 5.75 cm in length and 2.25 cm in width (Fig. 3.32).

The blind end or tip of the pouch is free of muscle, although the aperature is surrounded with a sphincter-like arrangement of muscle, and longitudinal muscle fibers extend into the pouch for almost two-thirds of its length. The wall of the pouch is 0.4–0.5 mm thick. The thickness of the various layers are 42 μm of stratified squamous epithelium, 42 μm of dense connective tissue, 182 μm of the muscle layer, and 149 μm of loose areolar connective tissue. There are no hairs, no glands, and no lymphatic vessels in this extremely vascular membrane.

The arterial blood supply is derived from the external carotid artery which branches to form the external maxillary. This vessel gives off the superior saccu-

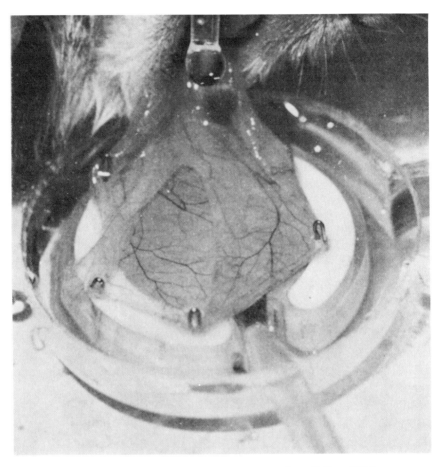

Fig. 3.32. Single layered pouch preparation. (From Duling, 1973.)

lar artery and the inferior labial artery. The inferior labial artery further divides to form the inferior and middle saccular arteries. The posterior saccular artery, which supplies blood to the posterior half of the pouch, has branches directly from the external carotid artery (Fig. 3.33).

The superior saccular artery serves the anterior dorsal quarter of the pouch wall. The inferior saccular artery branches to supply small arteries to the ventromedial surface of the pouch, and the middle saccular artery goes to the medial surface, dividing into small arteries that run parallel to muscle fibers (Hoffman et al. 1968).

Examination of the everted pouch stretched across a mount gives the impression of a rather random distribution of long, straight arterial vessels and parallel large veins that have numerous branches and form many arcades or interconnections.

VI. Special Tissues

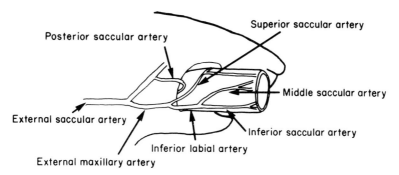

Fig. 3.33. Arterial blood supply to the hamster cheek pouch. (Modified from Priddy and Brodie, 1948.)

Low power magnification of a single layer preparation of the pouch confirms the impression of a profuse and random distribution of arterial vessels with numerous diminutive branches or offshoots and a greater number of large venous vessels which seem to dominate the vascular bed (Fig. 3.34).

At higher magnifications ($\geq 450 \times$), it is possible to determine the diameters of arterial vessels branching from the first order parent vessel down to those of the terminal arteriole (which is the fourth order of branching) and the capillary. The average diameter of a major artery extending into the most distal portion of the tip was found to be 76.7 μm, based on measurements in ten prepared pouches. The branches of the major artery, designated as small arteries or second order vessels, had an average diameter of 29.8 μm, which was less than half the diameter of their parent vessels. The arterioles or third order vessels had an average diameter of 15.2 μm, with a range between 9.0 and 22.5 μm. The wide variation in diameter of these vessels is probably due in part to differences in resting tonus of the encircling vascular smooth muscle. These vessels are remarkably sensitive to external mechanical stimuli, which frequently produces dilation, indicating high resting tone. The terminal arterioles or fourth order vessels showed an average diameter of 7.5 μm, although one-third of the vessels had diameters of 4.5 μm, indicating that perhaps identification between terminal arterioles and capillary vessels is not easy to determine. No tests for contractile activity of these smallest vessels were made, although such tests would have been a better indication of whether the vessel was a muscular arteriole with its precapillary sphincter or a capillary devoid of smooth muscle. Capillary vessels had an average diameter of 4.5 μm (Fig. 3.35).

It has been noted by investigators using this preparation that spontaneous and intermittent flow is produced by the action of sphincters located at the end of the arterial distribution and that arterioles between 24 and 28 μm exhibit spontaneous (phasic) vasomotion. It is reasonable to presume that this activity varies from normal with the type and depth of anesthesia. Blood pressure, determined by the use of a pressure cuff on the pouch, was found to be about 90 Hg, and direct

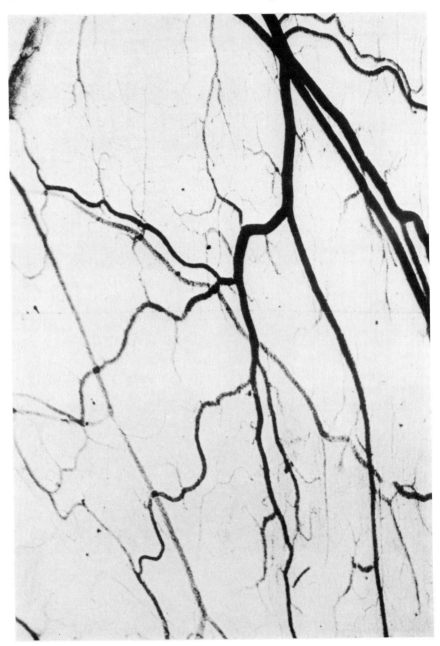

Fig. 3.34. Vascular pattern of the hamster cheek pouch. (From Wiedeman, 1963.)

VI. Special Tissues   61

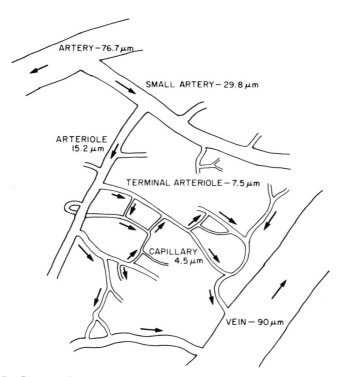

Fig. 3.35. Diagram of a typical pattern from artery to vein in the hamster cheek pouch.

measurement of systemic blood pressure from the carotid artery was 116 mm Hg. Numerous mast cells appear along the vessels as well as in the surrounding tissue.

The pouch, by comparison with most other sites used in studies of the microcirculation, is extremely vascular. Artery to artery and vein to vein anastomoses are prominent. It has not been possible to designate specific areas served exclusively by any of the numerous arteries. There is no arterial pathway that can be called a preferential channel through the capillary network such as described in the mesentery of dogs and rats.

The nerve supply to the vessels originates from the facial spinal accessory nerves. The distribution of the nerves is very diffuse and is difficult to trace into the pouch and to the smallest vessels. Through electron microscopic, autoradiographic, and histochemical techniques, adrenergic and cholinergic fibers have been demonstrated within walls of arterioles with diameters of 35–60 μm (first and second order vessels). Such dual innervation in these microvessels must be considered unique, since vasodilator nerves have not been demonstrated in the microcirculatory beds in other mammalian preparations.

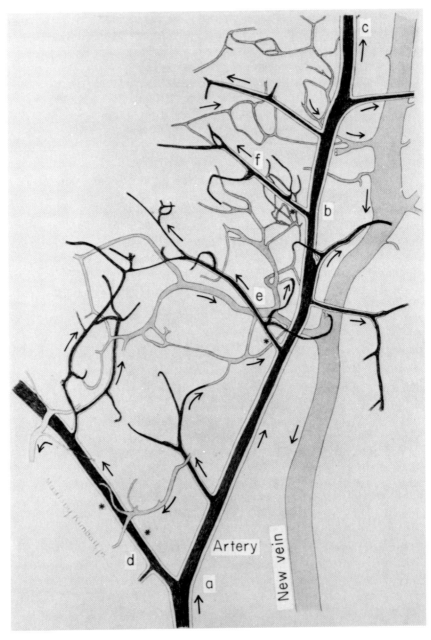

Fig. 3.36. Vascular pattern in the rabbit ear chamber. (From Clark and Clark, 1934.)

VI. Special Tissues

B. Rabbit Ear

The installation of a transparent chamber in the rabbit ear marks the beginning of microscopic observation at high magnification of blood vessels in the living mammal. The early papers by Sandison (1928) and by Clark and Clark (1932, 1934) contain most of the available descriptive material relative to the vascular patterns as well as characteristic behavior of the arterial, capillary, and venous vessels seen in the 1.5 cm chamber. Two types of chambers were used for microscopic observation. One was the preformed tissue chamber in which existing vessels were made visible by removing one layer of skin, and the other was the round table chamber in which new vessels grew into a vacant space. A camera lucida drawing of a portion of the vasculature in a chamber is shown in Fig. 3.36.

Four branching orders of arterial vessels can be seen before the capillary is reached, and the confluence of postcapillary vessels to form the venous outflow pathways is no different from that seen in other beds. Only capillary diameters are mentioned, and these range between 4.25 and 12.6 μm. The vascular pattern, showing arterioles, capillaries, and venules, is seen in more detail in Fig. 3.37.

Differences in the type of blood flow in these vessels can often be used in their identification. Arteries and arterioles have a steady, rapid flow, in contrast to the steady but slow flow of venous vessels, and capillaries have a flow that may be intermittent with frequent changes in direction.

Lymphatic vessels are prominent in the rabbit ear and constitute a specific system of vessels that run for long distances in close proximity, primarily with venous vessels. The wall of the lymphatic may actually appear to touch the wall of the vein. When the rabbit ear is heated and blood flow is increased, the lymphatics become more prominent. The same increase in the caliber of lymphatic vessels is noted during inflammation or infection in the chamber. Blood cells are seen in lymphatic vessels following a slight injury from pressure exerted on the cover slip of the chamber. Leukocytes are seen to migrate from venous to lymphatic vessels in certain conditions, and occasionally red blood cells are seen, especially after hemorrhage in the chamber.

The lymphatic endothelium is thinner and therefore more delicate than that of blood vessels. The lymphatic vessels are easily compressed by vasomotion of adjacent arterial vessels. It should be noted that lymphatic "capillaries" are often as large as the vein that they accompany and larger than the paired artery.

A prominent feature of the vasculature of the rabbit ear is the presence of arteriovenous anastomoses. These large connections between the arterial and venous vessels number about 40 in a chamber with an area of 1.5 cm. The diameter of an a-v anastomosis is usually in the range of 20-35 μm, and it may be straight, curved, or coiled (Fig. 3.38).

Spontaneous rhythmic contractions are characteristic of arterial vessels in the rabbit ear chamber. The rate is one to three times a minute in a resting, unanes-

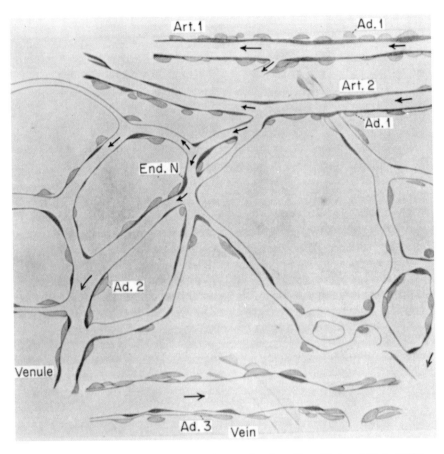

Fig. 3.37. Detail of microvessels in rabbit ear chamber. (From Clark and Clark, 1940.)

thetized rabbit (Clark and Clark, 1932). The contractions are visible in the main artery of the intact ear and are observed in arterial branches and the smallest arterioles seen in the chamber. Contraction of a precapillary arteriole (sphincter) alters or may stop flow in the capillaries distal to it. Different arterial branches and portions of the same branch contract at different times, seemingly independent of one another. Contractions are more frequent in the smaller arterial branches than in the main artery and its larger branches.

Arteriovenous (a-v) anastomoses also contract actively and may be the most actively contracting vessels in the rabbit ear; some contract as many as six to ten times a minute. Although no temperature regulating function had been assigned to them, it was recognized that the a-v anastomoses were more prominent and

VI. Special Tissues

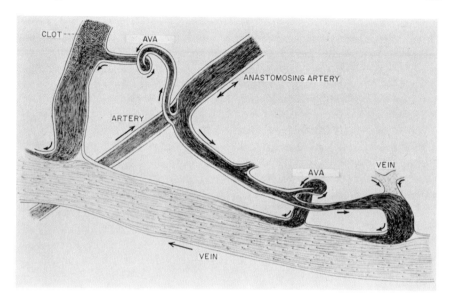

Fig. 3.38. Diagram of three arteriovenous anastomoses (AVA), all branches of the same artery, entering branches of the same vein. (From Clark and Clark, 1934.)

seemed more numerous on warm than on cold days, and that heating the ear causes tightly closed a-v anastomoses to open and become actively contractile.

The rabbit ear chamber has the advantage of use in chronic experiments or lengthy experiments of observation up to 6 months or more. In addition, the animals are docile and can be trained to lie quietly so the vessels can be observed in the unanesthetized state.

C. Human Skin

Using a titanium chamber installed in a twin pedicle tubed skin-flap on the inside of the upper arm in human volunteers, Branemark (1971) has studied the behavior of blood cells in microvessels. The tissue, observed at high magnification, is ordinary subcutaneous connective tissue with a microvascular topography and physiology equivalent to normal tissue. The microvascular flow pattern, vasomotion, and reactions are considered to be the same as in vessels of the tissue outside the chamber.

Arterioles larger than 20 μm show rapid flow with rhythmic changes that occur irregularly, sometimes appearing every few seconds and sometimes several minutes apart. Venules of equivalent size show similar changes in flow rate and the occasional adherence of leukocytes. Some leukocytes roll slowly and continuously along the vessel wall. Capillaries and arteriovenous shunts also have rhythmic changes in flow, probably dependent on upstream changes. The flow rate, expressed in min/sec, is given as 0.84 for arterioles, 0.52 for precapillary vessels and a-v shunts, 0.06 for capillaries, and 0.23 for venules. Asano *et al.* (1973) state that microcirculation of man under basal conditions is characteristic and very similar to that observed in the rabbit ear chamber.

Erythrocytes are easily deformed and can pass through vessels or intraluminal spaces as small as 1.5–2 μm. Diapedesis occurs easily. The red cells in passage can be described as assuming an umbrella shape (Fig. 3.39).

Crenated red cells appear in normal circulation and form amorphous masses during stasis. The red cells also form rouleaux in low flow states as well as stasis, but resume their normal shape as soon as flow is reinstituted.

White blood cells are much more rigid than erythrocytes, and red blood cells change shape when forced to squeeze past adherent ones. The white blood cells may adhere to the vessel wall or roll slowly along the wall, apparently not influenced by the velocity of the flow in the vessel. Adherent leukocytes may produce partial or complete obstruction to flow of blood in small vessels. These cells can emigrate through a vessel wall in 20–30 seconds, assuming a position adjacent to the outer wall of the vessel.

Platelets are seen as thin, rigid discs that vary in diameter size from 3 to 6 μm. They appear as separate cells in normal flow, adhering to red blood cells or to the vessel wall only after local tissue injury. At injury sites, the platelets will also adhere to one another to form a true platelet aggregate or microthrombus.

No characteristic vascular pattern has been described for the vessels inside the chamber.

D. Bat Wing

The extended wing of the little brown bat, *Myotis lucifugus,* measures 11 cm in length from the bat's shoulder to the fingers and 4 cm from the bone of the arm

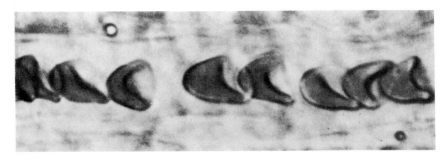

Fig. 3.39. Photograph of red blood cells in the vascular bed of a man as seen in the Branemark chamber. (From Branemark, 1971.)

VI. Special Tissues

to the bottom edge. In this rectangle of tissue there are two major arteries originating at the shoulder that course to the terminal margin; the smaller one is close to the body wall and the larger one transects the wing to divide the space into two triangles. The portion of the wing most frequently used for observation is the area between the large major artery and the lower margin. The major artery, with an average diameter of 97 μm, gives rise to approximately 13 vessels that branch at 90°–45° and continue toward the borders, where they divide into dichotomous branches that form arcades with similar branches from parallel arteries. In this vascular bed, branches from the major artery are called first order vessels. They have an average diameter of 52 μm and give rise to 12 branches, called second order vessels or small arteries, which average 19 μm in diameter. From these branches, 9.7 third order arterioles, 7 μm in diameter, leave at right angles. The final arterial distribution is the fourth order or terminal arteriole. These vessels are 5 μm in diameter and give rise to three 3 μm-diameter capillaries. An average of 4.6 terminal arterioles are formed from the third order arterioles.

Terminal arterioles are so named because they form no connections with any other arterial vessels. They are the parent vessels for the capillaries, which are interconnected to form a network that eventually becomes confluent, thus forming postcapillary venules. From this point, venous vessels and arterial vessels are paired. The dimensions and average number of branches are given in Table 3.2. A diagram of the vascular bed is shown in Fig. 3.40, and a detailed sketch of flow paths in a small area of the bat's wing is shown in Fig. 3.41.

TABLE 3.2
Dimensions of Blood Vessels in the Bat's Wing[a]

Vessel	Average length (mm)	Average diameter (μm)	Average number of branches	Number of vessels	Total cross-sec area (μm^2)	Capacity (mm$^3 \times 10^{-3}$)	Capacity (%)
Artery	17.0	52.6	12.3	1	2,263[b]	38.4	10.1
Small artery	3.5	19.0	9.7	12.3	4,144	14.4	3.8
Arteriole	0.95	7.0	4.6	119.3	5,101	4.7	1.2
Capillary	0.23	3.7	3.1[c]	548.7	6,548	1.5	0.39
Post-capillary venule	0.21	7.3		1,727.0	78,233	16.4	4.3
Venule	1.0	21.0	5.0	345.4	127,995	127.9	33.7
Small vein	3.4	37.0	14.1	24.5	27,885	94.7	25.0
Vein	16.6	76.2	24.5	1	4,882	81.0	21.4

[a] From Wiedeman, 1963a.
[b] Average of individual cross-sectional areas.
[c] Calculated.

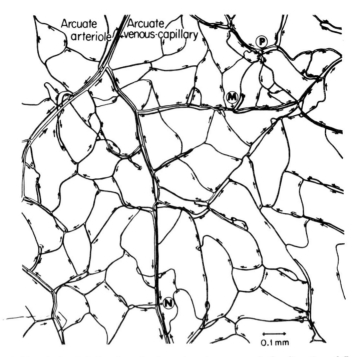

Fig. 3.40. A detailed drawing of a bat wing. Arrows mark the direction of flow. Areas marked M,N, and P indicate where flow from opposite directions meet and exit into a terminal arteriole. (From Webb and Nicoll, 1954.)

Information about the characteristics of this vascular bed is primarily from studies by Nicoll and Webb (1946, 1955; Nicoll, 1964, 1966); from their observations, numerous ideas arose regarding the regulation and distribution of flow in capillaries. They suggested that (1) the arcades or arcuate pattern serve to assure adequate blood supply for capillary networks, even though single arterial vessels may have wide fluctuations in flow; (2) the right angle formed by a branch from a parent vessel is a factor in producing an abrupt reduction in pressure so that equal pressure can be maintained in capillaries regardless of their distance from the arterial supply; (3) the muscular elements along arterioles respond myogenically to aid regulation of flow to more distal portions; (4) neural control of larger arteries is not important in capillary blood flow regulation; and (6) active vasomotion in terminal arterioles ultimately controls the flow in capillaries.

Spontaneously active contraction is seen in the arteriolar vessels, in veins, and in the lymphatic vessels of the bat wing. Also, variations in diameters of arterial vessels are very noticeable when the unanesthetized bat is attempting to move against the restraints imposed by the holder and the clips. This is probably a result of changes in sympathetic vasoconstrictor outflow. The smaller arteriolar

VI. Special Tissues

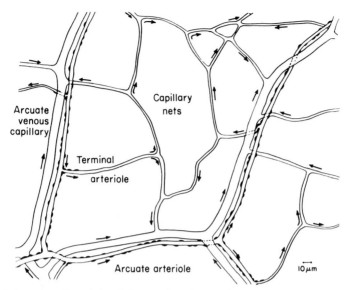

Fig. 3.41. An enlarged detail showing distribution of capillary vessels from a terminal arteriole and their confluence. (From Nicoll and Webb, 1955.)

vessels respond to myogenic stimuli when upstream arterial vessels are relaxed. Venous vasomotion is prominent and its rate and vigor vary with alterations in arterial flow.

Large nerve trunks are easily visualized as are their ramifications, which follow the branching of arterial vessels until the terminal portions of the bed are reached. Electrical stimulation of a larger nerve trunk results in the constriction of innervated vessels but also in contraction of skeletal muscle bands at the same frequency as that of the electrical stimulus. It causes liberation of a liquid from the glandular structures that surround the hair follicles. The skeletal muscle response indicates the presence of somatic motor nerves in the nerve trunk.

Spontaneous contractile activity of the lymphatic vessels, which are highly visible adjacent to the larger veins, is independent of central nervous control. It is enhanced with changes in volume of the intravascular compartment and with changes in vascular permeability.

No thoroughfare or preferential channels are seen in the vascular pattern of the bat wing in that any pathway between the arterioles and the venules can and does actively contract to complete closure from time to time. Muscular investment of arteriole vessels is not limited to sites or origin of capillary branches, but may continue along a terminal arteriole from which three or more capillaries form. These capillaries are distal to the final portion of the terminal arteriole, which is encircled by a single smooth muscle cell. This single smooth muscle cell is without question the precapillary sphincter in this bed.

VII. CONCLUSION

This survey of the composition of numerous vascular beds makes it apparent that there are a variety of patterns formed by the terminal blood vessels dictated primarily by the function of the tissues that they serve. Although there is a lack of uniformity in details for each bed, the descriptive material of the various tissues and organs that have been used for *in vivo* microscopy are consolidated here.

The immediate usefulness will be to the investigator who has selected a vascular bed for study and wishes to become familiar with its architecture and flow characteristics. In a sense, the chapter can be considered as a field guide to the vasculature of numerous tissues and organs. The long range usefulness has several facets. The information needed to complete the anatomical and functional characteristics of some of the beds is identified and can be supplied by investigators. Comparisons of the different vascular patterns may make similarities more recognizable and foster the use of the same terminology for blood vessels in various beds that share similar functions. An array of vascular patterns may emerge that can be divided into useful categories based on the type of tissue in which they appear. For example, do highly metabolically active tissues share similar features such as size of vessels, outflow pathways, arteriovenous anastomoses, or the number of capillaries originating from a parent arteriole?

To be able to review the descriptions of the vascular beds in a compact form may generate a cohesiveness between investigators who use different tissues so that flow characteristics discovered in one bed may be sought in another bed. For some time, vasomotion, or spontaneous contractile activity, was referred to primarily as a feature of bat wing vessels, but this alternating rhythmical contraction and relaxation of arterial vessels has now assumed an important role in regulation of flow into exchange vessels of numerous vascular beds. There must be many such instances where structural and functional features in one vascular bed are applicable to another.

REFERENCES

Asano, M., Branemark, P.-I., and Castenholz, A. (1973). A comparative study of continuous qualitative and quantitative analysis of microcirculation in man. *Adv. Microcirc.* **5**, 1–31.

Baez, S. (1961). Response characteristics of perfused microvessels to pressure and vasoactive stimuli. *Angiology* **12**, 452–461.

Baez, S. (1973). An open cremaster muscle preparation for the study of blood vessels by *in vivo* microscopy. *Microvasc. Res.,* **5** 384–394.

Baez, S. (1977). Skeletal muscle and gastrointestinal microvascular morphology. *In* "Microcirculation". G. Kaley and B. M. Altura, (eds.), Vol. 1, pp. 69–74. Univ. Park Press, Baltimore, Maryland.

Bassingthwaighte, J. B., Yipintsoi, T., and Harvey, R. B. (1974). Microvasculature of the dog left ventricular myocardium. *Microvasc. Res.,* **7**, 229–249.

References

Bessou P., and Laporte, Y. (1965). Technique de preparation d'une fibre afferente I et d'une fibre afferent II innervant le meme friseau neuro-musculaire, chez le chat. *J. Physiol. (Paris)* **57**, 511-520

Bloch, E. H. (1956). Microscopic observations of the circulating blood in the bulbar conjunctiva in man and health and disease. *Ergeb. Anat. Entwicklungsgesch.* **35**, 1-98.

Bohlen, H.G., and Gore, R.W. (1976). Microvascular pressures in innervated rat intestinal muscle and mucosa. *In* "Microcirculation" (J. Grayson and W. Zingg, ed.), Vol. 1, pp. 325-326. Plenum, New York.

Bohlen, H.G., and Gore, R.W. (1977). Comparison of microvascular pressures and diameters in the innervated and denervated rat intestine. *Microvasc. Res.*, **14**, 251-264.

Bohlen, H.G., Hutchins, P.M., Rapela, C.E., and Green, H.D. (1975). Microvascular control in the intestinal mucosa of normal and hemorrhaged rats. *Am. J. Physiol.* **229**, 1159-1164.

Branemark P.-I. (1971). "Intravascular Anatomy of Blood Cells in Man." Karger, Basel.

Burkel, W.E. (1970). The fine structure of the terminal branches of the hepatic arterial system of the rat. *Anat. Rec.* **167**, 329-350.

Burton, K.S., and Johnson, P.C. (1972). Reactive hyperemia in individual capillaries of skeletal muscle. *Am. J. Physiol.* **223**, 517-524.

Chambers, R., and Zweifach, B.W. (1944). Topography and function of the mesenteric capillary circulation. *Am. J. Anat.* **75**, 173-205.

Clark, E.R., and Clark, E.L. (1932). Observations on living preformed blood vessels as seen in a transparent chamber inserted into the rabbit's ear. *Am. J. Anat.* **49**, 441-477.

Clark, E. R., and Clark, E. L. (1934). The new formation of arteriovenos anastomoses in the rabbit's ear. *Am. J. Anat.* **55**, 407-467.

Clark, E.R., and Clark, E.L. (1940). Microscopic observations on the extra-endothelial cells of living mammalian blood vessels. *Am. J. Anat.* **66**, 1-49.

Davis, E., and Landau, J. (1966). "Clinical Capillary Microscopy." Thomas, Springfield, Illinois.

Ditzel, J., and Heuer, H.E. (1975). Caliber changes in the cutaneous microvessels as related to changes in basal metabolism in diabetics. *Bibl. Anat.* **13**, 219- 221.

Duling, B. R. (1973). The preparation and use of the hamster cheek pouch for studies of the microcirculation. *Microvasc. Res.* **5**, 423-429.

Eriksson, E., and Myrhage, R. (1972). Microvascular dimensions and blood flow in skeletal muscle. *Acta Physiol. Scand.* **86**, 211-222.

Fenton, B.M., Zweifach, B.W., and Worthen, D.M. (1979). Quantitative morphometry of conjunctival microcirculation in diabetes mellitus. *Microvasc. Res.* **18**, 153-166.

Forbes, H.S. (1928). The cerebral circulation I. Observation and measurement of pial vessels. *Arch. Neurol. Psychiatry.* **19**, 751-761.

Frasher, W.G., Jr. and Wayland, H. (1972). A repeating modular organization of the microcirculation of cat mesentery. *Microvasc. Res.* **4**, 62-76.

Fronek, K., and Zweifach, B.W. (1975). Microvascular pressure distribution in skeletal muscle and the effect of vasodilation. *Am. J. Physiol.* **228**, 791-796.

Fronek, K., and Zweifach, B. W. (1977). Microvascular blood flow in cat tenuissimus muscle. *Microvasc. Res.*, **14**, 181-189.

Fung, Y.C., and Sobin, S.S. (1969). Theory of sheet flow in the lung alveoli. *J. Appl. Physiol.* **26**, 472-488.

Gore, R.W., and Bohlen, H.G. (1975). Pressure regulation in the microcirculation. *Fed. Proc., Fed. Am. Soc. Exp. Biol.* **34**, 2031-2037.

Grafflin, A.L., and Corddry, E.G. (1953). Studies of peripheral blood vascular beds in the bulbar conjunctiva of man. *Johns Hopkins Hosp. Bull.* **93**, 275-289.

Grant, R.T. (1966). The affects of denervation on skeletal muscle blood vessels (rat cremaster). *J. Anat.* **2**, 305-316.

Gray, S.D. (1971). Effect of hypertonicity on vascular dimensions in skeletal muscle. *Microvasc. Res.*, **3**, 117–124.

Grayson, J., Davidson, J.W., Fitzgerald-Fitz, A., and Scott, C. (1974). The functional morphology of the coronary microcirculation in the dog. *Microvasc. Res.* **8**, 20–43.

Guth, P.H. (1977). The gastric microcirulation and gastric mucosal blood flow under normal and pathological conditions. *Prog. Gastroenterol.* **3**, 323–347.

Guth, P. H., and Rosenberg, A. (1972). In vivo microscopy of the gastric microcirculation. *Am. J. Dig. Dis.* **17**, 391–398.

Hammersen, F. (1968). The pattern of the terminal vascular bed and the ultrastructure of capillaries in skeletal muscle. *In* "Oxygen Transport in Blood and Tissue." (D. W. Lubbers, U. C. Luft, G. Thews, and E. Witzleb, eds.), pp. 184–197. Thieme, Stuttgart.

Hammersen, F. (1970). The terminal vascular bed in skeletal muscle with special regard to the problem of shunts. *In* "Capillary Permeability" (C. Crone and N. A. Lassen, eds.). Academic Press, New York, pp.351-371.

Henquell, L., LaCelle, P. L., and Hoñig, C. R. (1976). Capillary diameter in rat heart *in situ:* Relation to erythrocyte deformability, O_2 transport, and transmural O_2 gradients. *Microvasc. Res.* **12**, 259–274.

Henrich, H. N., and Hecke, A. (1978). A gracilis muscle preparation for quantitative microcirculatory studies in the rat. *Microvasc. Res.* **15**, 349–356.

Hoffman, R.A., Robinson, P.F., and Magalhaes, H. (1968). "The Golden Hamster." Iowa State Univ. Press, Ames.

Hutchins, P.M., Bond, R.F., and Green, H.D. (1974). response of skeletal muscle arterioles to common carotid occlusion. *Microvasc. Res.* **7**, 321–325.

Irwin, J. W., and MacDonald, J. (1953). Microscopic observations of the intrahepatic circulation of living guinea pigs. *Anat. Rec.* **117**, 1-15.

Irwin, J. W., Burrage, W. J., Aimar, C. E., and Chesnut, R. W. (1959). Microscopic observations of the pulmonary arterioles, capillaries and venules of living guinea pigs and rabbits. *Anat. Rec.* **119**, 391–408.

Knisely, M. H. (1936). Spleen studies. I. Microscopic observations of the circulatory system of living unstimulated mammalian spleens. *Anat. Rec.* **65**, 23–50.

Lee, R. E., and Holze, E. A. (1950). The peripheral vascular system in the bulbar conjunctiva of young normotensive adults at rest. *J. Clin. Invest.* **29**, 146–150.

Lipowsky, H. H., and Zweifach, B. W. (1974). Network analysis of microcirculation of cat mesentery. *Microvasc. Res.* **7**, 73–83.

Ma, Y. P., Koo, A., Kwan, H. C., and Cheng, K. K. (1974). Online measurement of the dynamic velocity of erythrocytes in the cerebral microvessels in the rat. *Microvasc. Res.* **8**, 1–13.

McCuskey, R. S. (1966). A dynamic and static study of hepatic arterioles and hepatic sphincters. *Am. J. Anat.* **119**, 455–476.

McCuskey, R. S., and McCuskey, P. A. (1977). *In vivo* microscopy of the spleen. *Bibl. Anat.* **16**, 121–125.

Majno, G., Palade, G. E., and Schoefl, G. K. (1961). II. Studies on inflammation. The site of action of histamine and serotonin along the vascular tree: A topographic study. *J. Biophys. Biochem. Cytol.* **11**, 607–626.

Martini, J., and Honig, C. R. (1969). Direct measurement of intercapillary distance in beating rat heart *in situ* under various conditions of O_2 supply. *Microvasc. Res.* **1**, 244–256.

Myrhage, R., and Hudlicka, O. (1976). The microvascular bed and capillary surface area in rat extensor hallucis proprius muscle (EHP). *Microvasc. Res.* **11**, 315–323.

Navari, R. M., Wei, E. P., Kontos, H. A., and Patterson, J. L., Jr. (1978). Comparison of the open skull and cranial window preparations in the study of the cerebral microcirculation. *Microvasc. Res.* **16**, 304–315.

References

Nicoll, P. A. (1964). Structure and function of minute vessels in autoregulation. *Circ. Res.* **15**, 1245-1253.

Nicoll, P. A. (1966). Intrinsic regulation in the microcirculation based on direct pressure measurements. *In* "The Microcirculation, A Symposium" (W. L. Winters and A. N. Brest, eds.) Thomas, Springfield, Illinois. 89-102.

Nicoll, P. A., and Webb, R. L. (1946). Blood circulation in the subcutaneous tissue of the living bat's wing. *Ann. N.Y. Acad. Sci.* **46**, 697-711.

Nicoll, P. A., and Webb, R. L. (1955). Vascular patterns and active vasomotion as determiners of flow through minute vessels. *Angiology* **6**, 291-308.

Phillips, S. J., and Rosenberg, A., Meir-Levi, D., and Pappas, E. (1979). Visualization of the coronary microvascular bed by light and scanning electron microscopy and x-ray in the mammalian heart. *Scanning Electron Microsc.* **3**, 735-742.

Plyley, M. J., Sutherland, G. J., and Groom, A. C. (1976). Geometry of the capillary network in skeletal muscle. *Microvasc. Res.* **11**, 161-173.

Priddy, R. B., and Brodie, A. F. (1948). Facial musculature, nerves and blood vessels of the hamster in relation to the cheek pouch. *J. Morphol.* **83**, 149-180.

Rappaport, A. M., Knisely, M. H., Ridout, J. H., and Best, C. H. (1958). Microcirculatory changes in the liver of choline-deficient rats. *Proc. Soc. Exp. Biol. Med.* **57**, 522-524.

Rappaport, A. M. (1973). The microcirculatory hepatic unit. *Microvasc. Res.* **6**, 212-228.

Reilly, F. D., and McCuskey, R. S. (1977). Studies of hemopoietic microenvironment VI. Regulatory mechanisms in the splenic microvascular system of mice. *Microvasc. Res.* **13**, 79-90.

Rhodin, J. A. G. (1978). Microscopic anatomy of the pulmonary vascular bed in the cat lung. *Microvasc. Res.* **15**, 169-193.

Rosenberg, A., and Guth, P. H. (1970). A method for the *in vivo* study of the gastric microcirculation. *Microvasc. Res.* **2**, 111-112.

Rosenblum, W. I. (1965). Cerebral microcirculation: A review emphasizing the interrelationship of local blood flow and neuronal function. *Angiology* **16**, 485-507.

Rosenblum, W. I., and Zweifach, B. W. (1963). Cerebral microcirculation in the mouse brain. *Arch. Neurol.* **9**, 414-423.

Sandison, J. C. (1928). The transparent chamber of the rabbit's ear, as seen with the microscope. *Am. J. Anat.* **41**, 447-474.

Smaje, L., Zweifach, B. W., and Intaglietta, M. (1970). Micropressure and capillary filtration coefficients in single vessels of the cremaster muscle of the rat. *Microvasc. Res.* **2**, 96-110.

Sobin, S. S., and Tremer, H. M. (1972). Diameter of myocardial capillaries. *Microvasc. Res.* **4**, 330. (Abstr.)

Sobin, S. S., Tremer, H. M., and Fung, Y. C. (1970). Morphometric basis of the sheet flow concept of the pulmonary alveolar microcirculation in the cat. *Circ. Res.* **26**, 397-414.

Tillmanns, H., Ikeda, S., Hansen, H., Sarma, J. S. M., Fauvel, J.-M., and Bing, R. J. (1974). Microcirculation in the ventricle of the dog and turtle. *Circ. Res.* **34**, 561-569.

Tuma, R. F., Childs, M., Intaglietta, M., and Arfors, K.-E. (1975). Microvascular flow pattern in the tenuissimus muscle. *Bibl. Anat.* **13**, 151-152.

Wearn, J. T., Ernstene, A. C., Bromer, A. W., Barr, J. S., German, W. J., and Zschiesche, L. J. (1934). The normal behavior of the pulmonary blood vessels with observations on the intermittence of the flow of blood in the arterioles and capillaries. *Am. J. Physiol.* **109**, 226-256.

Webb, R. L., and Nicoll, P. A. (1954). The bat wing as a subject for studies in homeostasis of capillary beds. *Anat. Rec.* **120**, 253-265.

Wiedeman, M. P. (1963a). Dimensions of blood vessels from distributing artery to collecting vein. *Circ. Res.* **12**, 375-378.

Wiedeman, M. P. (1963b). Patterns of the arteriovenous pathways. *In* "Handbook of Physiology, Circulation II" (W. F. Hamilton and P. Dow, eds.), Sect. 2, Vol. II, pp. 891-933. Am. Physiol. Soc., Washington, D.C.

Zweifach, B. W. (1954). Direct observation of the mesenteric circulation in experimental animals. *Anat. Rec.* **120**, pp. 277-288.

Zweifach, B. W. (1974a). Quantitative studies of microcirculatory structure and function: 1. Analysis of pressure distribution in the terminal vascular bed in cat mesentery. *Circ. Res.* **34**, 843-857.

Zweifach, B. W. (1974b). Quantitative studies of microcirculatory structure and function: II. Direct measurement of capillary pressure in splanchnic mesenteric vessels. *Circ. Res.* **34**, 858-866.

Zweifach, B. W., and Lipowsky, H. H. (1977). Quantitative studies of microcirculatory structure and function: III Microvascular hemodynamics of cat mesentery and rabbit omentum. *Circ. Res.* **41**, 380-390.

Zweifach, B. W., and Metz, D. B. (1955). The terminal vascular bed of mesenteric structures and skeletal muscle. *Angiology* **6**, 282-289.

Zweifach, B. W., and Prather, J. W. (1975). Micromanipulation of pressure in terminal lymphatics in the mesentery. *Am. J. Physiol.* **228**, 1326-1335.

4

Methods of Preparation of Tissues for Microscopic Observation

I. INTRODUCTION

The selection of the site for microscopic observation of a vascular bed in a living animal is of primary importance in any experimental design. The vascular areas in current use show many variations in visible components, in geometrical patterns, and in the ease or difficulty of preparation. The vascular area to be used must satisfy the needs of the problem to be explored. The hamster cheek pouch, for example, would be an inappropriate bed if lymphatic behavior was a part of the protocol; studies concerned with velocity of blood flow might be more difficult in beds requiring direct rather than transmitted light; and the overall effects of anesthesia and extensive surgery are always factors that must be considered. If drugs or other agents are required directly at the point of observation, the selected bed should have vessels suitable for intravenous or intra-arterial cannulation or for topical application.

A review of the numerous preparations in use, with a discussion of some of the known advantages and disadvantages, should be helpful in choosing the bed most suitable for a given set of experiments.

At the onset, the investigator should be aware of the effects of various anesthetic agents on the behavior of blood vessels and, if necessary for the success of the study, utilize a preparation designed for the unanesthetized human or animal.

Most anesthetics will produce variations from normal in the cardiovascular system. The most commonly used anesthetics for small laboratory animals are pentobarbital, chlorolose, and occasionally ether. Morphine may be used in conjunction with other anesthetics for special purposes.

Alterations in blood vessel diameters depends in part on the depth of anesthesia. Under light anesthesia arterial blood vessels may be more constricted than usual, whereas deep anesthesia produces vasodilation of these vessels. Deep anesthesia can also cause a reduction in systemic blood pressure with a resultant decrease in velocity of blood flow. In some instances, an anesthetic can increase heart rate well beyond normal values and thus affect cardiac output. Catecholamine release is promoted by some anesthetics.

It has also been noted that spontaneous contractile activity, or vasomotion, is depressed by anesthetics. When the tonus of vessels is reduced and the spontaneous regulatory activity of precapillary vessels is diminished, a change in capillary permeability and fluid exchange may result.

An additional effect of anesthesia may be seen as a result of depression of the respiratory center caused by most barbiturates. Respiration may be reduced to such an extent that acidosis with increased P_{CO_2} and reduced oxygenation of the blood may occur.

Body temperature of an anesthetized animal is generally reduced, which may lead to the appearance of an excessive number of leukocytes in the circulating blood.

A large number of leukocytes adhering to or rolling along vessels walls is considered by some investigators as an indication of an abnormal condition that should be corrected before proceeding with observations. Leukocytes circulating in excess of normal numbers can effectively change blood flow characteristics in the terminal vasculature.

An important consideration in the maintenance of normal flow in exteriorized tissues is the composition and temperature of the suffusing fluid that bathes exposed tissue. It should simulate the usual environment of the tissue so that it does not alter the behavior of the vascular smooth muscle or its responsiveness to internal controls or external stimuli. It is recommended that a means to test responsiveness of vessels be devised to monitor the condition of the vascular components throughout experimental procedures.

It is obvious that an awareness of the effects of anesthetics is essential in assessing the condition of a vascular bed used for studies of the microvasculature. Careful consideration must be given to the selection and dosage of the agents used for surgical preparation and maintenance of the animal during observation and experimental procedure.

A comment about microscopic equipment for visualization of vascular beds may be helpful before selecting the particular animal and tissue to be used for study. A good optical system is of major importance and so is an appropriate light. For preparations in which small animals such as rats and hamsters are used,

II. Exteriorization of Internal Tissues

a standard research microscope is adequate. For larger animals in which tissues or organs are observed *in situ*, modifications of standard equipment are usually required.

There are four primary procedures for viewing terminal vascular beds in the living animal using transmitted light. (1) Internal tissues may be brought outside the body to permit transmission of light for microscopic observation. (2) Transparent chambers may be implanted in selected areas. (3) Tissues or organs may be examined *in situ*. (4) Naturally occurring superficial structures that are transparent enough may be used. Each of the procedures has it advocates because of certain inherent advantages.

These four basic methods will be presented with a general description after which the specific preparations will be briefly discussed. References are given so that the interested reader can review the current literature for exact details pertaining to the preparation of his choice.

II. EXTERIORIZATION OF INTERNAL TISSUES

There are some tissues of animals that can be extended from their normal position in the body and spread over specially constructed mounts to permit microscopic observation. Tissues such as mesentery and omentum are thin enough to allow transmission of light and strong enough to withstand the necessary surgical procedures required to position them outside the body without adversely affecting blood flow. In addition to these tissues, which need only to be pulled to the outside through an abdominal incision, certain muscles, such as the cremaster and the spinotrapezius, are thin and translucent and thus lend themselves to visualization of their blood vessels after surgical preparation. A third type of tissue is the cheek pouch of the hamster, which can be everted and, after some dissection, spread over a pedestal for viewing.

The exposed tissues can be protected by keeping them moist with a physiological saline solution, or by covering them with mineral oil, Saran Wrap, or a glass cover slip. The external environment may be kept at body temperature by using a heated suffusing solution, or, if the tissue is covered, by heating the mounting on which the tissue rests.

A. Mesentery of the Rat and Cat

The surgery and preparation of the rat cecal mesentery for microscopic observation as described by Zweifach (1954) is relatively simple. Rats weighing between 100 and 150 gm are the most satisfactory. They are anesthetized with an intraperitoneal injection of a selected anesthetic agent, usually a barbiturate, and placed in a prone position for cannulation of the trachea and the femoral vein.

After these procedures are completed, a small transverse incision is made in the right side of the abdomen just distal to the last rib. The cecum of the rat is very large and can be palpated with gentle pressure. It is helpful to manipulate the cecum to the right side of the abdominal cavity before the abdominal incision is made. The cecum can be pushed through the incision to the outside after which it is covered with moist cotton. The rat is turned to the right and the large, horseshoe-shaped piece of intestine held in semicircular position by mesenteric tissue is arranged around a glass or plastic pedestal, so that the mesentery is loosely stretched over the surface of the pedestal. Suffusion of the exposed tissue with warm Ringer's gelatin solution is begun immediately. A more recent description by Zweifach was published in 1973 (Zweifach, 1973). Zweifach considers that the structural and metabolic simplicity of the bed make it a good site for selected studies.

Other mesenteric preparations use a portion of the regular mesentery tissue, which requires that a larger segment of the small bowel be exteriorized than needed for visualization of the rat cecal mesentery. Currently popular in some laboratories is the cat mesentery. According to the method described by Frasher and Wayland (1972), the peritoneal cavity of the cat is opened by a midline incision in the area of the umbilicus. The small bowel is pulled out and placed in a heated chamber filled with a suitable physiological solution. After succinylcholine chloride, the animal is ventilated by positive pressure. A selected portion of mesentery is exposed and restrained by two or three padded clamps at the antimesenteric border. A method described by Zweifach (1973) is simpler in that the mesenteric tissue is placed loosely over a hollow plastic pedestal and the intestine covered with moistened cotton. No attempt is made to stretch the mesentery, and the animal breathes normally. Care should be taken not to stretch the tissue that connects the mesentery to the body wall, in order to avoid a reflex vasodilation. If temperature and respiration are normal, little or no peristalsis develops. The preparation should last without change for several hours, and the first sign of deterioration may be leukocyte adhesion in venules, later followed by extravasation.

Several criteria must be met before the bed is considered suitable for use. An indication of good vascular tone is seen when the large arteries crossing the mesentery are not excessively dilated. In general, their diameter should be about one-third that of the accompanying vein. Blood flow through small vessels should be rapid and capillary flow usually shows intermittentcy with no regularity.

B. Gastric Circulation

To make microscopic observations of the gastric circulation in the living animal, Rosenberg and Guth (1970) developed two preparations, one to visualize the serosal, muscular, and submucosal circulation in the stomach, and one to

II. Exteriorization of Internal Tissues

observe the superficial mucosal circulation. Fasted rats, 150–200 gm, are anesthetized with sodium pentobarbital given intraperitoneally. After performing a tracheostomy, succinyl chloride is given, and the animal is placed on a small animal respirator. To exteriorize the stomach, a midline epigastric incision is made, gastrohepatic ligaments are cut, and the stomach is exteriorized manually. It is covered with gauze pads and continously irrigated with Ringer's gelatin solution heated at 37°C. A small incision is made in the duodenum through which a quartz rod light carrier is passed into the stomach. The pencil of light from the quartz rod can be directed as desired by a prism attached at the end of the carrier. With this arrangement it is possible to observe the flow of blood in the serosal and muscular layers of the stomach wall. Submucosal circulation is seen best after dissection of the serosal and muscular layers away from a small area.

To observe the mucosal circulation, it is necessary to make an inverted pouch so that the mucosa is exteriorized. Using a magnification of 200× or more, blood flow in capillaries of the superficial mucosa is visible.

C. Rat Intestinal Muscle and Mucosa

Bohlen and Gore (1976) introduced their description of the preparation of intestinal muscle and mucosa by pointing out that the microcirculation of the tissues is clearly visible and directly accessible, and therefore one can perform a wide variety of quantitative measurements.

While applicable to several animals, these investigators have used the albino rat, anesthetized with an intraperitoneal injection of 10% urethane and 2% chloralose, 0.6 ml/100 gm body weight. The rat is placed on a lucite microscopic stage and body temperature maintained with a heating mat. The trachea is cannulated, as is the left femoral artery, for monitoring systemic blood pressure. An incision of 1 cm is made in the abdomen on the right side and a loop of intestine is drawn through with a cotton tipped applicator and placed on a gauze pad. The exposed tissue is suffused constantly with physiological saline at 37°C, a point which is emphasized by the authors. A longitudinal incision of about 1.5 cm is made through a section of the intestine wall along the antimesenteric border, and the slit section of intestine is draped over a pedestal. To observe blood vessels of the intestinal muscle, the section is arranged with the muscle surface facing up; however, the mucosa must be facing the microscope objective if mucosal villi are to be examined (Fig. 4.1). Many important details are given to help avoid problems with the preparation that might be encountered by an investigator first adopting this technique, e.g., how to control unwanted intestinal motility with isoproterenol, how to control blood flow in and out of the observed area, and how to surgically denervate the vessels. The preparation lasts for 6 hours or more and several criteria can be used to determine if the bed is still viable. The appearance of white blood cells sticking to the walls, a decrease in arteriolar diameter, an increased vasomotion with an irregular frequency, and the

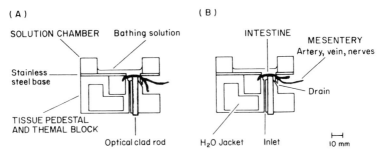

Fig. 4.1. The solution chamber and pedestal for microscopic observation of intestinal muscle (A) and mucosa (B). (From Bohlen and Gore, 1976.)

cessation of mucosal secretion from the mucosa indicate deterioration at which point the experiment should be terminated.

D. Hamster Cheek Pouch

The hamster cheek pouch preparation was first introduced in 1946 (Fulton *et al.*, 1946) and has been modified by successive investigators since then. Most preparations currently follow the basic instructions of Duling (1973) with modifications as necessitated by the particular purpose of the experiment. In general, after anesthesia, the animal is secured on a small table or board in a supine position for the insertion of a tracheal cannula and a femoral vein cannula for supportive injections of anesthesia or other soluble agents as required. A cheek pouch, usually the left one but dependent on the design of the board, is everted with a cotton applicator and anchored by its tip to the outer margin of a circular opening in the board. The open space in the board permits transmission of light from below and usually contains a pedestal surrounded by a moat. The pouch is spread across the entire surface of the opening and pinned to an encircling ring of compliant material such as rubber. This double layer of tissue reveals a profuse vascular pattern. The visibility of the vessels can be greatly enhanced by sectioning the double layer through the midline of the upper layer and reflecting the margins of this layer to the sides where the tissue is pinned to the rubber ring. The remaining thin, single layer of the pouch must be cleared of connective tissue by careful dissection to further increase visibility. The exposed tissue can be suffused with an appropriate physiological solution maintained at a constant temperature or protected from drying with a cover slip or mineral oil. Duling recommends several tests for the viability or reactivity of the preparation such as (1) cessation of vasomotion, (2) venular stasis, (3) loss of arteriolar tonus, and (4) arterial stasis, all of which indicate an abnormal condition.

The bed is very vascular and offers a display of major arterial vessels through

II. Exteriorization of Internal Tissues

capillaries back to veins. Magnifications as high as 900–1200× can be used. A preparation will remain useable for several hours.

E. Cremaster

Several technical improvements have been introduced in the preparation of the rat cremaster muscle since it was first used by Majno *et al.* (1961). Grant (1964) further developed the technique, and 3 years later Majno *et al.* (1967) refined their original method by inserting a white spatula as a light-reflecting surface directly behind the muscle. It was necessary to inject gallamine triethiodide (Flaxedil) to keep the muscle motionless, and this adds the necessity of artificial respiration for the rat. Baez (1973) prepared an open cremaster muscle, which overcame some of the disadvantages of the earlier methods. Figure 4.2 shows the sequence of the procedures developed by Baez.

The rat is placed in a supine position and the scrotum is extended by means of

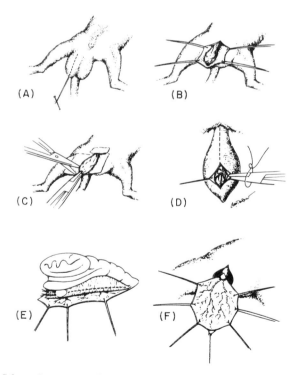

Fig. 4.2. Schematic sequence illustrating preparation of the rat cremaster muscle for microscopic observation. (From Baez, 1973.)

a thread placed at the tip. An incision is made in the midline of the ventral surface of the scrotum and then connective tissue fascia is gently dissected from around the cremaster sack. Vessels between the testis and epididymis can be tied or severed with a cautery after which the testis can be extirpated or pushed back into the abdomen. The cremaster muscle is then spread over an appropriate mount. Maintenance of proper temperature (a heated mount may be used) and an appropriate suffusing fluid should prevent fasciculation. The exposed cremaster muscle remains attached to the rat's body by a small pedicle which contains the cremaster artery and vein and the nerves that innervate the tissue.

III. TRANSPARENT CHAMBERS

Transparent chambers implanted in suitable tissues have the advantages of being most useful for chronic studies and allowing the investigator to make microscopic observations in an unanesthetized animal.

The first chambers were developed for use in the rabbit ear by Sandison (1924) and Clark *et al.* (1930) at the Wistar Institute in Philadelphia. Numerous modifications and improvements have been made since then. For detailed description, the reader is referred to an article by Nims and Irwin (1973).

In recent years, transparent chambers have been successfully implanted in the skin of rats (Algire, 1954) and in the human upper arm by Branemark (1971) in Sweden. The latest site, used for transplantation of tissues from other parts of the body, is a chamber in the hamster cheek pouch (Greenblatt *et al.*, 1969).

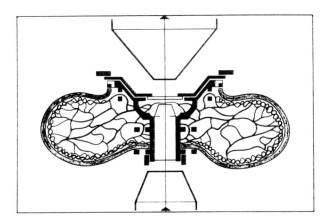

Fig. 4.3. A transverse section of the chamber, the skin tube, and the microscope ready for observation. (From Branemark, 1971.)

III. Transparent Chambers

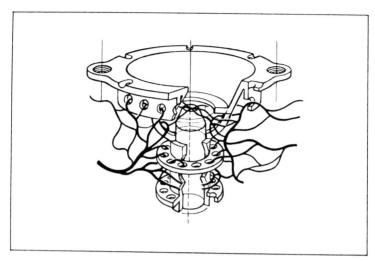

Fig. 4.4. The pattern formed by vessels growing into the chamber. (From Branemark, 1971.)

A. Transparent Chambers in Man

The visualization of human microvessels in a transparent chamber was developed by Per-Ingvar Branemark over a period of about 10 years (Branemark, 1971). A thin layer of vascularized tissue grows into a twin pedicled skin tube on the inside of the upper arm of a human volunteer. The tissue is enclosed between a glass cover slip and a quartz glass rod (Fig. 4.3).

A twin pedicled, tubed skin flap is prepared by surgery on the inside of the left upper arm and allowed to heal for 3 months. After this, the titanium chamber containing a light-conducting quartz rod and cover slip is installed in the tube of skin. In the 50μm distance between the cover slip on the upper surface and the quartz rod on the lower surface, blood vessels and tissue grow into the space. After 3–4 weeks, the tissue appears as ordinary subcutaneous connective tissue with microvessels considered to be similar to those in surrounding normal tissue (Fig. 4.4).

The chamber causes no discomfort and can be viewed with transmitted light up to 8 months after installation. See Fig. 4.5 for appearance of chamber in the skin.

B. Rabbit Ear Chamber

The observation of blood vessels in a transparent chamber inserted into the rabbit ear was first introduced by Sandison (1924) and used by Clark and Clark (1932) has continued as a useful site to this day.

84 4. Tissue Preparation for Microscopic Observation

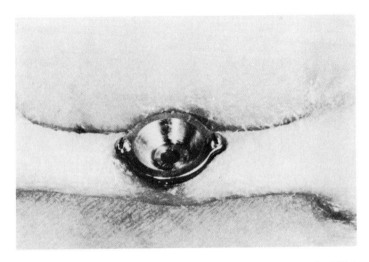

Fig. 4.5. The skin tube with a chamber in place. (From Branemark, 1971.)

Basically, the method consists of fastening the two parts of a chamber, the top and bottom, to the cartilage of the rabbit ear. An area in the ear may be prepared in such a way that a new vascular system will grow in between the two portions of the fixed chamber, or a "preformed" chamber may be used in which the original tissues are viewed after removal of cartilage and skin of the inner side of the ear. Other designs permit introduction of drugs or chemical agents in studies of vascular reactions.

A successful chamber implantation, free of infection or debris, permits clear observations of blood vessels for a long period of time in an unanesthetized mammal.

C. Hamster Cheek Pouch Chamber

The method for preparing a chamber in the hamster cheek pouch was introduced by Sanders and Shubik (1964), who felt that it had better optical qualities than the rabbit ear chamber and was ideally suited for the transplantation of tissues. Numerous modifications were made after the original description and a detailed account is given in a paper by Greenblatt *et al.* (1969).

Precautions are taken to prevent subsequent infection from developing in the implanted chamber. The hamsters are given Terramycin in their drinking water before the surgery and postoperatively for the duration of the experiment, a practice which reduces the infection rate from 34 to 5%. Sterile procedures are used throughout the surgery. A portion of the chamber, called the baseplate, is

IV. In Situ Tissues and Organs

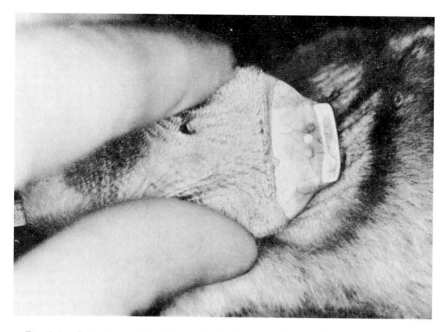

Fig. 4.6. Retraction of the skin anteriorly from an incision at the posterior part of the baseplate. (From Greenblatt et al., 1969.)

inserted into the pouch after which the overlying skin is retracted to expose the cheek pouch membrane (Fig. 4.6). The mucoareolar layer is then removed by gentle dissection to improve visibility after which a top plate is put in place over the membrane. If transplanted tissue is to be used, it is put in place before the pouch is covered by the top plate (Fig. 4.7). Observation of the vasculature enclosed in the chamber is made using long working distance objectives, and transillumination is achieved by inserting a light-conducting rod into the animal's mouth (Fig. 4.8).

This innovative technique is currently being used for tissue transplant studies such as renal hemograft transplants, and it has been successfully demonstrated that other tissues such as heart and lung will develop in the chamber. Its limitations have not been established.

IV. IN SITU TISSUES AND ORGANS

Transillumination was the major problem to overcome in observing the microcirculation of living organs in their normal positions. The introduction of the

4. Tissue Preparation for Microscopic Observation

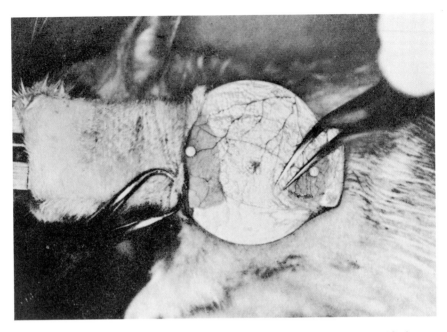

Fig. 4.7. A transplant placed on the dissected membrane before covering with the top plate. (From Greenblatt et al., 1969.)

quartz rod technique by Knisely (1936) solved the problem, and today the "light pipe" is used in a number of preparations of organs and tissues *in situ*. Light is conducted by internal reflection from one end of the rod to the other, so the light source can be some distance away from the tissue. The rod can be bent to any desired shape so it can be placed conveniently under the tissue. The usual precautions are taken for maintaining a suitable environment for the exposed tissue.

A. Tenuissimus Muscle

The tenuissimus muscle as a site for *in vivo* microscopic observation of skeletal muscle vasculature was introduced by Branemark and Eriksson (1972). The rabbit or cat are used as experimental animals, anesthetized with either chloralose or urethane given intravenously. A longitudinal incision is made in the skin of the dorsolateral part of the thigh, the subcutaneous fascia is carefully dissected, and the major part of the tenuissimus muscle is exposed. Because the nerves and vessels which supply the muscle are at the medial portion, they are not damaged in the surgical procedure. The muscle is left in place, and a special condenser is placed underneath it. Tuma *et al.* (1977) used the lateral saphenous

IV. *In Situ* Tissues and Organs

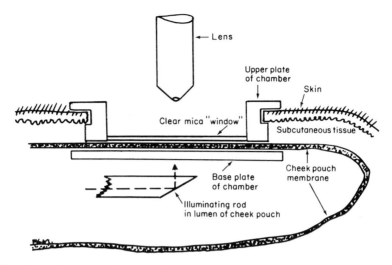

Fig. 4.8. Diagram to illustrate illumination of the chamber with a light pipe. (From Greenblatt *et al.*, 1969.)

vein as a landmark for the initial incision since it runs parallel to the teniussimus muscle. After reflecting the quadriceps muscle to expose the tenuissimus, a space is made to insert a fiberoptic rod underneath the muscle. Krebs-Henseleit solution flows continuously and a water immersion lens is used for observation. A few other modifications of the original technique are described by Fronek and Zweifach (1975).

B. Spleen

The method described here for visualization of the spleen is that of McCuskey *et al.* (1972; McCuskey and McCuskey, 1977). It is applicable to the mouse, rat, and hamster. An intraperitoneal injection of either urethane or sodium pentobarbital is recommended as the anesthetic agent. The tip of the spleen is brought to the outside through a 1 cm incision just under the rib cage on the left side, and it is placed over a mica window in the microscope stage. The spleen is covered on both the upper and lower surface with Saran Wrap attached to a movable frame. The Saran Wrap serves to hold the organ in position and limits movements due to respiration and the beating heart, but it has no rigidity and therefore does not compress the blood vessels in the tissue (Fig. 4.9).

The tissue is bathed in warm Ringer's solution. By the combination of eyepieces and objectives, magnifications of 200–1350× can be achieved. McCuskey recommends transillumination with light with a wavelength between 575 and 750 nm to eliminate absorption of light by the hemoglobin of the red

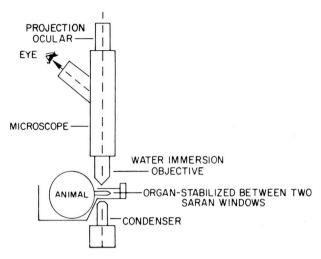

Fig. 4.9. Diagram of optical equipment used for *in vivo* microscopic study of the spleen. (From McCuskey and McCuskey, 1977.)

blood cells. With this method it is possible to differentiate between red and white pulp, to identify arterioles, sinusoids, and venules, to see the flow of blood in the vessels, to watch the individual cells flowing through the sinusoids, and even to observe cytoplasmic and nuclear details of cells developing outside of the blood vessels.

C. Pial Circulation

Pial circulation is studied in the living animal through the microscope in either an open preparation in which the exposed cortex is covered only with a suitable solution or in a closed preparation in which the cortex is covered by a glass window. The cranial window technique is used for both acute and chronic experiments, but obviously makes chronic experiments possible.

Cats, rabbits, and rats are most frequently used when craniotomies are performed. The pial circulation can be observed, however, through the relatively thin, translucent skull of some small animals.

The initial procedures are essentially the same with all animals. The animals are anesthetized, a tracheotomy tube is inserted, skeletal muscle relaxation and artificial ventilation is often used to reduce movement, and the animal's head is placed in a head holder. A midsaggital incision is made, and the scalp is reflected to the sides. After the bone surface is cleaned, a trephine is used to make a circular hole in the skull. Artificial cerebrospinal fluid heated to 37°C or heated

IV. *In Situ* Tissues and Organs

mineral oil is used to cover the exposed brain. The dura mater is cut and reflected toward the edges of the circular opening. The exposed vessels can be continuously suffused with an appropriate solution or can be covered with a slip held in a frame which serves the dual purpose of providing an optically flat surface and reducing evaporation. This type of chamber is described by Ma *et al.* (1974). A more complex cranial window preparation, which can be implanted and used for chronic experiments, is described in detail by Levasseur *et al.* (1975). This preparation has allowed these investigators to observe the pial circulation in an unanesthetized rabbit for prolonged periods of time in excess of a year. The design of the cranial window with valves permits flushing of the exposed pial vessels, which is a distinct advantage over other implanted chambers, and is believed to be the reason for its long period of usefulness.

D. Lung

The preparation of the lung for observation through the microscope has been reported by Wearn *et al.* (1934) and by Irwin *et al.* (1954). They differ in that in one case the lung is left in place in the chest and observed through a window, whereas the second method involves bringing a portion of the lung to the outside.

Wearn and co-workers (1934) viewed the lung in cats by dissecting away muscle between the eighth and ninth ribs until only the parietal pleura remained. The pleura appeared as a clear, transparent, intact membrane, 0.6–1.2 cm in diameter. The lower edge of the lower lobe of the lung could be seen. A midline abdominal incision was made to create another window in the diaphragm by dissecting muscle from the abdominal surface until the parietal pleura was exposed from this angle. The tip of the lung was between the window in the chest wall and the window in the diaphragm. A beam of light was directed through a quartz rod to the window in the diaphragm so that the lung was transilluminated and pulmonary vessels could be observed through a microscope placed over the window in the chest wall. Observations were made at magnifications of 90–117×.

In some studies the cats breathed normally; in others curare was used to abolish respiratory movements and the lungs were inflated through the trachea.

The method of Irwin *et al.* (1954) eliminated respiratory movements by intratracheal insufflation of oxygen, given at a rate of 1.5 liters/min in rabbits and 0.5 liters/min in guinea pigs. A skin incision was made at the level of the fifth rib, and thoracic wall muscles were ligated and cut after which intercostal muscles were sectioned. The ribs were separated, the pleura incised, and a portion of lung was brought through the opening. The lung was kept moist with warmed Ringer's solution, and the edge of the exposed lung was transilluminated for observation. Transillumination was accomplished by placing a quartz rod under the edge of the lung. Observations were made at 48–500×.

E. Liver

The method used by McCuskey (1966) for exposure of rat liver for observation is presented here.

Rats are anesthetized intravenously with sodium pentobarbital, and a tracheal cannula inserted. The liver was exposed by making an L-shaped abdominal incision that extended along the linea alba and then laterally along the right subcostal margin to the twelfth rib. The falciform ligament was cut to partially free the liver from the diaphragm to reduce movement of the liver caused by respiration. Also, skeletal muscle paralysis was produced with succinylcholine, and intratracheal oxygen was administered. To reduce movements of the liver caused by the heart and the gastrointestinal system, the liver was immobilized on the mechanical stage of the microscope by positioning a lobe of the liver over a window covered with Saran Wrap and securing the lobe with a movable clamp. Gauze was placed between the liver and the intestines to further reduce movement caused by intestinal movement.

The liver will remain in good condition for observation on an average of 4 hours.

F. Coronary Circulation

Microscopic observation of blood vessels in some portion of most tissues has not been difficult to achieve once the problems of illumination and preservation of a normal environment have been solved. Major difficulties were encountered, however, in organs exhibiting rhythmical movements such as the heart and the lungs. Martini and Honig (1969) introduced a method for observing blood vessels in a beating heart about $20 \mu m$ below the epicardial surface.

Rats, 170–240 gm, were prepared for observation by opening the chest and removing the pericardium to expose the relatively flat surface of the right ventricular free wall for viewing. The rats were ventilated at a constant rate with a respiratory pump. A microscopic objective with a working distance great enough to prevent contact between the microscope and the beating heart was selected, and a point source strobe light which was triggered by the film transport of a movie camera permitted the photographing of the heart. The microscope was focused on the heart in its position at the end of expiration, a time when respiratory movement stopped briefly. For each 100–150 ft of film, approximately 50 frames were in focus. These focused frames were then projected, having a final magnification of 500–1000×, and measurements were made from the projected image. Subsequently, this laboratory added some modifications (Henquell *et al.*, 1976) and stated that the most important limitation of the method was that only capillaries within 20 μm of the epicardial surface could be observed for measurement of their diameters; therefore, the measurements did not apply to capillary diameters during systole that were below this depth.

A second *in vivo* method is that of Tillmanns *et al.* (1974) who inserted a 20

V. Superficial Structures

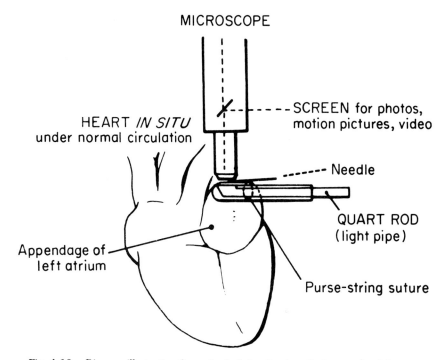

Fig. 4.10. Diagram illustrating the method of visualization of microvessels of the atrium using a needle containing a quartz rod which is inserted into the muscle. (From Tillmanns *et al.*, 1974, by permission of the American Heart Association, Inc.)

gauge light-transmitting needle just underneath the superficial layer of the myocardium to achieve transillumination (Fig. 4.10).

A system was developed to maintain the focal distance between the beating heart and the stationary objective of the microscope by having the objective of the microscope move in unison with the cardiac surface. Microcinematographs were taken on 16 mm color films at a rate of 400 frames/sec. When projected for viewing, the optical magnification was 2400×. The rate of filming made it possible to track a red blood cell in transit frame by frame.

Although there are limitations to visualization of the coronary circulation in both of these methods, they have made it possible to see some of the characteristics of the microcirculation.

V. SUPERFICIAL STRUCTURES

The *in vivo* microscopic observation of small blood vessels in superficial structures began shortly after compound lens were found to have the power of magnification. In the earliest studies the tails of tadpoles, fins of sticklebacks, and the web

92 4. Tissue Preparation for Microscopic Observation

between frog's toes were favorite sites of observation. Since amphibian preparations have been replaced by mammalian preparations, the remaining superficial tissue most used is the wing of the bat.

Bat Wing

One of the very few mammalian preparations in use which permits observation of the microvasculature in an unanesthetized animal is the wing of the bat. The method was developed by Nicoll and Webb (1946) and has been in use since then by several investigators. A detailed report of the preparation was written by Wiedeman (1967) and later in 1973 (Wiedeman, 1973).

This is perhaps the least complicated of all of the current preparations, requiring no anesthesia because no surgical procedures are involved. The bats, usually collected from their natural habitat by the investigator, can be kept in a state of hibernation in an ordinary refrigerator. The dormant bat is removed from the refrigerator and placed face down on a microscope slide (2″ × 6″) which is bounded by plastic (Fig. 4.11). The body of the bat is gently tucked into a plastic

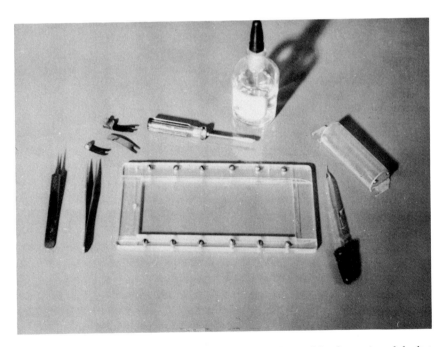

Fig. 4.11. Plastic bordered microscope slide and holder used in observation of the bat wing.

V. Superficial Structures 93

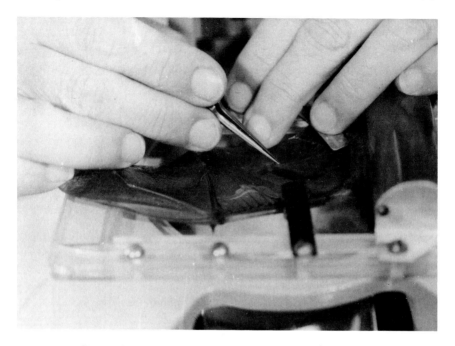

Fig. 4.12. A bat wing in position for microscopic observation.

tubular holder and one or both wings are extended across the surface of the slide and held in place with spring clips (Fig. 4.12).

The wing can be made more translucent by placing a drop or two of mineral oil between the undersurface of the wing and the microscope slide. When very high magnification is required, the upper layer of epithelial cells can be removed by teasing away the tissue with fine forceps and then covering the exposed vessels with physiological saline and a cover slip which is held off of the denuded area by slivers of glass placed at each side. The bat, a nocturnal animal, is content to lie quietly throughout the day. The intact wing shows no change in vascular activity or flow at any time, whereas the denuded preparation is subject to the same deterioration with time as other exposed vessels. Leukocyte adherence and the subsequent extravasation of leukocytes is one of the cardinal signs of change, as well as the cessation of spontaneous contractile activity of arteriolar vessels. The vascular bed can be easily denervated and also perfused through both the arterial and venous vessels. The extremely thin wing, about $20 \mu m$ excluding the areas of large vessels, results in highly visible structures with good resolution and great clarity.

Wiederhielm and Weston (1973) have used the Mexican freetail bat in the unanesthetized state to obtain pressure recordings. The bats were subjected to training sessions during which they were kept in a chamber for several hours with one wing extended and held with cotton applicators. Both wing surfaces were covered with mineral oil to reduce mobility of the wing. The animals did not indicate any stress from the restraint and low heart rates and blood pressures indicated that they often were asleep.

VI. CONCLUSION

This chapter is intended to assist the investigator in the selection of a microvascular bed appropriate to his needs. It is also intended to reduce the time it takes for an investigator to become expert in preparing the particular observation site that he has chosen. Knowledge of the changes that can be brought about by anesthetics and by improper maintenance of exteriorized tissues can be circumvented and therefore should expedite the successful preparation of a near-normal portion of living tissue for observation of its vasculature. There is a limit, however, to what can be learned from the written word, and the best learning will come from actual preparation of the tissue. In this way the surgical instruments best suited to the preparer, the accessories that should be in easy reach, and the unexpected difficulties in surgery, in cannulation, in suturing, in post-operative care and in maintenance of the preparation can be discovered. It is also helpful to visit the laboratory of someone who has mastered the technique. In this way, many technical procedures that may seem too simple to be mentioned in a written description or that may go unrecognized as a source of difficulty can be observed and imitated.

REFERENCES

Algire, G. H. (1954). The transparent chamber technique for observation of the peripheral circulation, as studied in mice. *In* "Peripheral Circulation in Man" (G. E. W. Wolstenholme and J. S. Freeman, eds.), Ciba Foundation Symposium, pp. 56–63. Little, Brown, Boston, Massachusetts.

Baez, S. (1961). Response characteristics of perfused microvessels to pressure and vasoactive stimuli. *Angiology* **12,** 452–461.

Baez, S. (1973). An open cremaster muscle preparation for the study of blood vessels by *in vivo* microscopy. *Microvasc. Res.* **5,** 384–394.

Bohlen, H. G., and Gore, R. W. (1976). Preparation of rat intestinal muscle and mucosa for quantitative microcirculatory studies. *Microvas. Res.* **11,** 103–110.

Branemark, P. I. (1971). "Intravascular Anatomy of Blood Cells in Man," Monograph. Karger, Basel.

Branemark, P. I., and Eriksson, E. (1972). Method for studying qualitative and quantitative changes of blood flow in skeletal muscle. *Acta Physiol. Scand.* **84,** 284–288.

Clark, E. R., and Clark, E. L. (1932). Observations on living performed blood vessels as seen in a transparent chamber inserted into the rabbit's ear. *Am. J. Anat.* **49,** 441–477.

References

Clark, E. R., Kirby-Smith, H. T., Rex, R. O., and Williams, R. G. (1930). Recent modifications of the method of studying living cells and tissues in transparent chambers inserted in the rabbit's ear. *Anat. Rec.* **47,** 187-211.

Duling, B. R. (1973). The preparation and use of the hamster cheek pouch for studies of the microcirculation. *Microvasc. Res.* **5,** 423-429.

Frasher, W. G., Jr., and Wayland, H. (1972). A repeating modular organization of the microcirculation of cat mesentery. *Microvasc. Res.* **4,** 62-76.

Fronek, K., and Zweifach, B. W. (1975). Microvascular pressure distribution in skeletal muscle and the effect of vasodilation. *Am. J. Physiol.* **228,** 791-796.

Fulton, G. P., Jackson, R. G., and Lutz, B. R. (1946). Cinephotomicroscopy of normal blood circulation in the cheek pouch of the hamster, Cricetus auratus. *Anat. Rec.* **96,** 537.

Grant, R. T. (1964). Direct observation of skeletal muscle blood vessels (rat cremaster). *J. Physiol. (London)* **172,** 123-137.

Greenblatt, M., Choudari, K. V. R., Sanders, A. G., and Shubik, P. (1969). Mammalian microcirculation in the living animal: Methodologic considerations. *Microvasc. Res.* **1,** 420-432.

Henquell, L., LaCelle, P. L., and Honig, C. R. (1976). Capillary diameter in rat heart *in situ*: Relation to erythrocyte deformability, O_2 transport and transmural O_2 gradient. *Microvasc. Res.* **12,** 259-274.

Irwin, J. W., Burrage, W. S., Aimar, C. E., and Chestnut, R. W., Jr. (1954). Microscopical observations of the pulmonary arterioles, capillaries, and venules of living guinea pigs and rabbits. *Anat. Rec.* **119,** 391-408.

Knisely, M. H. (1936). A method of illuminating living structures for microscopic study. *Anat. Rec.* **64,** 499-523.

Levasseur, J. E., Wei, E. P., Raper, A. J., Kontos, H. A., and Patterson, J. L. (1975). Detail description of a cranial window technique for acute and chronic experiments. *Stroke* **6,** 308-317.

Ma, Y. P., Koo, A., Kwan, H. C., and Cheng, K. K. (1974). Online measurement of the dynamic velocity of erythrocytes in the cerebral microvessels in the rat. *Microvasc. Res.* **8,** 1-13.

McCuskey, R. S. (1966). A dynamic and static study of hepatic arterioles and hepatic sphincters. *Am. J. Anat.* **119,** 445-476.

McCuskey, R. S., and McCuskey, P. A. (1977). *In vivo* microscopy of the spleen. *Bibl. Anat.* **16,** 121-125.

McCuskey, R. S., Meineke, H. A., and Kaplan, S. M. (1972). Studies of the hemopoietic microenvironment. II. Effect of erythropoietin on the splenic microvasculature of polylythemic CF_1 mice. *Blood* **39,** 809-813.

Majno, G., Palade, G. E., and Schoefl, G. K. (1961). Studies on inflammation. The site of action of histamine and serotonin along the vascular tree: A topographic study. *J. Biophys. Biochem. Cytol.* **11,** 607-262.

Majno, G., Gilmore, V., and Leventhal, M. (1967). A technique for the microscopic study of blood vessels in living striated muscle (cremaster). *Circ. Res.* **21,** 823-832.

Martini, J., and Honig, C. R. (1969). Direct measurement of intercapillary distance in beating rat heart *in situ* under various conditions of O_2 supply. *Microvasc. Res.* **1,** 244-256.

Nicoll, P. A., and Webb, R. L. (1946). Blood circulation in the subcutaneous tissue of the living bat's wing. *Annals of New York Academy of Sciences* **46,** 697-711.

Nims, J. C., and Irwin, J. W. (1973). Chamber techniques to study the microvasculature. *Microvasc. Res.* **5,** 105-118.

Rosenberg, A., and Guth, P. H. (1970). A method for the *in vivo* study of the gastric microcirculation. *Microvasc. Res.* **2,** 111-112.

Sanders, A. G., and Shubik, P. (1964). A transparent window for use in the Syrian hamster. *Isr. J. Exp. Med.* **11,** pp. 118.

Sandison, J. C. (1924). A new method for the microscopic study of living growing tissues by the introduction of a transparent chamber in the rabbit's ear. *Anat. Rec.* **28,** 281-287.

Tillmanns, H., Skeda, S., Hansen, H., Sarma, J. S. M., Fauvel, J. M., and Bing, R. J. (1974). Microcirculation in the ventricle of the dog and turtle. *Circ. Res.* **34,** 561-569.

Tuma, R. F., Lindbom, L., and Arfors, K. E. (1977). Dependence of reactive hyperemia in skeletal muscle on oxygen tension. *Am. J. Physiol.* **233,** H289-H294.

Wearn, J. T., Ernstene, A. C., Bromer, A. W., Barr, J. S., German, W. J., and Aschieache, L. J. (1934). The normal behavior of the pulmonary blood vessels with denervations on the intermittence of the flow of blood in the arterioles and capillaries. *Am. J. Physiol.* **109,** 236-256.

Wiedeman, M. P. (1967). A preparation for microscopic observation of circulation in the unanesthetized animal. In "*In Vivo* Techniques in Histology" (G. H. Bourne, ed.), pp. 162-180. Williams and Wilkins, Baltimore, Maryland.

Wiedeman, M. P. (1973). Preparation of the bat wing for *in vivo* microscopcy. *Microvasc. Res.* **5,** 417-422.

Wiederhielm, C. A., and Weston, B. V. (1973). Microvascular, lymphatic, and tissue pressures in the unanesthetized mammal. *Am. J. Physiol.* **225,** 992-996.

Zweifach, B. W. (1954). Direct observation of the mesenteric circulation in experimental animals. *Anat. Rec.* **120,** 277-288.

Zweifach, B. W. (1973). Microcirculation. *Annu. Rev. Physiol.* **35,** 117-150.

REGULATION OF FLOW AND EXCHANGE

5

Factors Involved in the Regulation of Blood Flow

I. INTRODUCTION

The function of the circulatory system is to deliver blood to all the cells of the body to provide them with a constant environment. Thus, it supplies the cells with nutrient materials, circulates their humoral products, and removes their wastes. Because of the physical properties of the exchange process which permits these activities, all cells of the body must be in close proximity to the blood vessels, and therefore, the number of blood vessels must be quite large. If all of the vessels were filled with blood and had constant blood flow, the volume of blood would be tremendous and the energy needed to propel the blood would be excessive. The body has a number of mechanisms that make the process of blood distribution an efficient one. The circulatory system, which possesses a highly specialized architectural design, has the ability to vary the proportion of cardiac output received by any one organ at a given time. Redistribution of cardiac output is achieved through the integration of local and neural mechanisms. Local control, which occurs in the microcirculation, allows the matching of blood flow to tissue need. Factors which affect blood flow into the microvessels will be the subject of this chapter.

The factors that control the flow of blood through the microvasculature through changes in diameter of microvessels can be divided into four general categories; neural, humoral, metabolic, and myogenic. A discussion of each of these control mechanisms will be presented, followed by a discussion of specific actions in different vascular beds. Before entering into this discussion, it should be pointed out that the relative importance of each of these mechanisms in controlling blood flow to a tissue is determined somewhat by the function of the tissue.

II. CONTROL OF MICROCIRCULATION

A. Neural Control

The central nervous system controls the diameter of the larger blood vessels and the tonus (or tone) of the investing vascular smooth muscle primarily through sympathetic vasoconstrictor nerves. When considering the neural control of blood vessels it should be kept in mind that the major function of the nervous system in this case is to maintain systemic blood pressure by altering vascular resistance. Also, blood flow to less essential organs may be temporarily sacrificed through neural control in order to perfuse organs essential for survival of the body. This means that, in a given organ, a discrepancy between the need for blood and the amount of flow actually supplied may occur. In microcirculation one may picture two opposing control systems; neural control, which restricts blood flow to the tissues in order to maintain systemic blood pressure, and local control mechanisms, which attempt to maintain blood flow at the level necessary for optimal function of each of the organs. The predominent effect will depend on the intensity of stimulation of each controlling factor in the various organs.

The extent to which neural stimulation is able to vary blood flow by changing resistance varies from organ to organ. Morphological studies have demonstrated differences in the density of adrenergic nerve terminals in different vascular beds which indicate regional differences in sympathetic vasomotor control. Functional differences have also been demonstrated.

The vessels that are directly innervated by sympathetic nerves are the larger vessels, such as arteries, small arteries, and some arterioles, whereas the terminal arterioles with their precapillary sphincters are considered to be under local control. Rhodin (1967) has shown in ultrastructure studies that nerves are present at all levels of the arterial vascular distribution in the fascia of rabbit thigh muscles, and nerve terminals are seen close to the smallest vessels. Furness (1973) and Furness and Marshall (1974) have presented convincing evidence, both anatomical and physiological, that whereas large arteries, small arteries, and arterioles are innervated by a network of adrenergic fibers, and whereas all of these vessels respond to nerve stimulation by constricting, the smaller precapillary arterioles have only a few adrenergic nerve fibers near them, and they do not respond to nerve stimulation. See Fig. 5.1 for a diagram of nerve distribution.

II. Control of Microcirculation

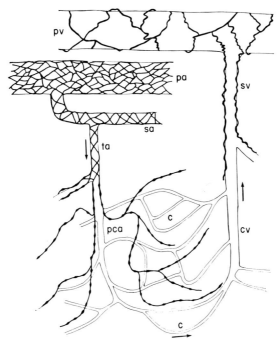

Fig. 5.1. Relationship between adrenergic nerves and the mesenteric blood vessels: pa, principal artery; pv, principal vein; sa, small artery of the microvasculature; ta, terminal arteriole; pca, precapillary arteriole; c, capillary; cv, collecting venule; sv, small vein. The adrenergic nerves are represented by the heavy lines, and arrows indicate the direction of blood flow. Note that the precapillary arterioles and the collecting venules are not innervated. (From Furness and Marshall, 1974.)

The rat intestinal mesentery was used for these experiments of Furness and Marshall in which microscopic observation was made of the arterial vessels during stimulation, and the presence of adrenergic nerves was established using the fluorescence histochemical method. Furness (1973) reviews other investigations that have also demonstrated the termination of adrenergic innervation before the final precapillary arterioles are reached, and he has shown that these non-innervated vessels are independent of a direct influence of sympathetic nerves. Furness states that his findings refute the suggestion that non-innervated vessels are affected by transmitter substances released from nerve terminals that are in close proximity. Other physiological evidence is found in bat wing studies by Wiedeman (1968), who showed that both chemical and surgical denervation resulted in relaxation of arteries, small arteries, and arterioles, whereas terminal arterioles and their parent vessel actually showed a decrease in diameter (Fig. 5.2). In addition to a smaller diameter, the vessels showed an increase in spontaneous contractile activity, which was presumed to have been a myogenic response

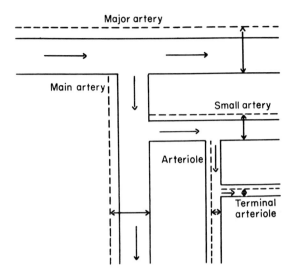

Fig. 5.2. Diagrammatic representation of the direction of diameter change of all vessels as a result of surgical denervation. The dashed lines show vascular diameter following denervation. (From Wiedeman, 1968.)

evoked by upstream vasodilation and increased blood flow to the more distal area. While this does not prove total lack of innervation of the terminal arterioles, it does show the dominance of local control and its effectiveness when upstream control is lacking.

The concept that larger arterial vessels are under sympathetic nervous system control while the final ramifications of the arterial system are under local control has a very important functional significance. This is especially true when one considers the need for separating the mechanisms that determine total peripheral resistance to support systemic blood pressure from the mechanisms that control distribution of blood flow in capillaries. Although the amount of blood flowing to a tissue would be reduced by the constriction of arterial vessels, the simultaneous dilation of terminal arterioles and precapillary sphincters could actually increase the number of capillaries containing flowing blood. An increase in the number of capillaries would at least partially compensate for the decrease in total flow so that the exchange of materials at the capillary level would be minimally affected.

The presence of vasodilator nerve fibers has been demonstrated in a number of tissues, but the physiological significance of the fibers has yet to be established. It should be kept in mind that vasoconstriction results from an increase in excitation of sympathetic nerves, and vasodilation results from a decrease in the discharge of impulses through these nerves. In most tissues, excluding the brain and the myocardium, sympathetic vasoconstrictor nerves have a tonic activity that results in a sustained contractile state of the vascular smooth muscle of innervated vessels, and vasodilation occurs when sympathetic nerve stimulation is

II. Control of Microcirculation

reduced. Vasodilator nerves, which would actively produce relaxation, are not the physiological mechanism that counteract sympathetic vasoconstrictor activity.

B. Metabolic Control

The responsiveness of vascular smooth muscle in the microcirculation to metabolites produced by the cells of the parenchymal tissue surrounding the blood vessels is an exquisite coupling between the needs of the tissue and the amount of blood delivered. An increase in the amount of metabolites can result from an increase in tissue activity or a decrease in blood flow to the site. As tissue metabolism increases, metabolic by-products accumulate around the precapillary vessels, causing them to dilate. The dilation of blood vessels promotes an increasing blood flow through the capillary network which results in an increased delivery of oxygen and a simultaneous removal of metabolic by-products. This will allow blood flow to equilibrate at a new level, matching the needs of the tissue. A similar situation is produced when the perfusion pressure to a tissue decreases. The resultant decrease in blood flow permits metabolic by-products to accumulate around the precapillary vessels with a resultant vasodilation, and blood flow again approaches normal levels. If the metabolic rate of a tissue decreases or if the perfusion pressure increases, then a lower concentration of vasodilator metabolites will be in the area, and precapillary vessels constrict thereby reducing blood flow to the capillary bed. This matching of blood flow to tissue activity makes the body capable of redistributing blood flow among organs. The metabolic control of vascular diameter enables the tissue to oppose the central nervous system control of blood flow to these beds.

Accumulation of metabolites or a change in tissue osmolarity will affect blood vessels, as shown in Table 5.1.

1. Carbon Dioxide

Carbon dioxide is a product of aerobic metabolism which has potent vasodilatory effects. There still remains a question as to whether CO_2 directly produces vasodilation or whether the hydrogen ion is the mediator for the vasodilator response. The mechanism may vary between tissues. In either case, an increase in tissue metabolism or a decrease in blood flow will cause CO_2 concentration to increase and pH to decrease. The importance of CO_2 and pH in the local regulation of blood flow varies between tissues. In the brain, for example, CO_2 may be the primary mechanism of control, whereas in the myocardium it may be of minimal importance.

2. Oxygen

The role of oxygen in controlling local blood flow has been the subject of debate for many years. Although it has been clearly demonstrated that decreasing oxygen delivery to the tissues results in a decrease in precapillary resistance, the

TABLE 5.1

Major Chemical Substances Regulating Blood Flow in the Microcirculation[a]

Agent	Response
Adenosine and adenine nucleotides	vdil, vcs
Hypoxemia	vdil
H^+	vdil
K^+	vdil
Inorganic phosphate	vdil
Hypercapnea	vdil
Kreb's cycle intermediates	vdil
Prostaglandins, thromboxanes, and prostacyclins	vcs, vdil, modifiers
GI tract polypeptides	
Glucagon	vdil
Cholecystokinin	vdil
Secretin	vdil
Pancreozymin	vdil
VIP	vdil
Hyperosmolarity	vdil

[a] Modified after Altura, 1978.

mechanism for this decrease is controversial. It has been demonstrated in isolated aortic strips and large arterial vessels that contractile tension is inversely proportional to oxygen availability (Carrier et al., 1964; Smith and Vane, 1966; Detar and Bohr, 1968). It has been suggested that these experiments demonstrate a direct effect on vascular tone. The ability of oxygen to directly affect vascular tone in normally functioning vascular beds depends on the oxygen tension being below the level necessary for maximal contractile activity by the smooth muscle in the resistance vessels. Decreases in oxygen tension within the physiological range can be demonstrated to cause dilation of large blood vessels and aortic strips. However, it is argued that this is a result of the barrier to oxygen diffusion that is inherent in the thickness of the large-walled vessels. According to Pittman and Duling (1973), smooth muscle cells in vessels the size of arterioles would not be affected by changes in oxygen tension at levels above 2–6 mm Hg. Strong support for this argument was presented by these investigators in a later study (Duling and Pittman, 1975). In this study, the hamster cheek pouch was suffused with a solution of varying oxygen tension. Micropipets were placed next to arterial microvessels. By surrounding the arterial vessels with a suffusion solution having an oxygen tension different from that of the solution suffusing the parenchymal tissue, it was shown that decreasing the oxygen tension (down to 2 mm Hg) around the microvessels did not result in vasodilation. However, an equivalent reduction in oxygen tension in the parenchymal tissue did result in a relaxation of these vessels. Therefore, these studies indicate that oxygen does not normally have a direct effect on the vascular tone in the arterial microvessels, but rather has an indirect effect through metabolites produced by the parenchymal tissue.

3. Adenosine

Adenosine is an example of a metabolic by-product which has a vasodilatory effect and whose concentration is influenced by oxygen tension. Adenosine has been proposed by Berne and his co-workers (Berne, 1963; Rubio and Berne, 1969) to be the agent primarily responsible for the local regulation of myocardial blood flow. It has also been implicated in the regulation of blood flow in skeletal muscle.

The mechanism of control of blood flow by adenosine is described as follows: A reduction in the cellular oxygen tension results in a breakdown of adenine nucleotides to form adenosine. Adenosine then diffuses down its concentration gradient into the interstitial fluid where it causes the smooth muscle cells in the arterial vessels to dilate. This relaxation increases blood flow and oxygen delivery to the tissue.

4. Potassium

Repetitive depolarization of parenchymal cells can cause an increase in the potassium concentration of the interstitial fluid. Potassium has been shown to be a very potent vasodilator, and therefore, the increase in interstitial concentration associated with an increase in activity may be one of the factors responsible for increasing blood flow. An increase in extracellular potassium concentration is thought to increase Na^+-K^+ ATPase activity causing an increase in vascular smooth muscle membrane polarity and subsequent relaxation of the smooth muscle. Potassium has been implicated in the local regulation of blood flow in the skeletal muscle and the brain.

5. Tissue Osmolarity

An increase of the interstitial fluid concentration of osmotically active solute, which could occur during repetitive contraction of skeletal muscle cells, results in a dilation of the blood vessels. Studies of isolated smooth muscle have shown that increasing the osmolarity of the surrounding solution hyperpolarizes the smooth muscle cells, which slows the rate of contraction and interferes with the spread of excitation between cells. Placing smooth muscle cells in a hyposmotic solution causes an increase in the activity of the smooth muscle cells. Changes in tissue osmolarity therefore offer another mechanism whereby increased tissue activity will cause an increase in tissue blood flow, and decreased activity will cause a reduction of flow.

6. Prostaglandins

Prostaglandins are extremely potent vasoactive compounds, which are produced in virtually every tissue of the body, including blood vessels. Arachadonic acid is the precursor of all prostaglandins. Prostaglandins may act on blood vessels directly, by modulation of sympathetic neurotransmitter release and uptake, or by modulation of hormonal action on blood vessels. With the exception of the pulmonary vasculature, prostaglandins of the E series cause dilation of blood vessels in all circulatory beds *in vivo*. The response of the vasculature to prostaglandins of the F series depends on the species, tissue, and dose. Recently PGI_2, a short-lived metabolite of the endoperoxides, has been shown to produce pronounced dilation of vascular smooth muscle. The physiological and pathological importance of the short-lived intermediates are difficult to evaluate, but this is currently an area of intensive investigation and may prove to be of great significance in the regulation of blood flow.

C. Myogenic Control

A third mechanism which may contribute to the regulation of blood flow through the microcirculation is myogenic control. Bayliss (1902) was the first to suggest that distention of blood vessels is actually a stimulus for the contraction of smooth muscle cells in the vessel wall. According to the myogenic hypothesis, the smooth muscle cells in the vessel wall are partially constricted in response to distention, even under conditions when there is no innervation or exposure to blood-borne vasoactive agents. Since the vasculature would normally be in a state of partial constriction, vascular resistance would be decreased as the vessels relax when distention decreases, or be increased in response to an increase in transmural pressure. In both cases the myogenic responsiveness of the blood vessels would tend to keep blood flow constant.

It has been known for many years that other types of smooth muscle, e.g., intestinal, respond myogenically to stretch. The cells contract spontaneously, have an unstable resting membrane potential, and conduct depolarization between cells. These electrophysiological properties have also been demonstrated in small pre- and postcapillary vessels, showing that the vascular smooth muscle cells have the potential for myogenic responsiveness (Funaki, 1961).

Numerous microcirculatory studies have been made in an attempt to demonstrate the myogenic behavior of blood vessels in the microcirculation. By injecting saline intra-arterially into the bat wing, Wiedeman (1966) showed that the increase in intraluminal pressure was associated with an increase in the contractile activity of the arterioles, extending the time during which blood flow was occluded by vasomotion of the arterioles. Johnson and Wayland (1967) also demonstrated the influence of intraluminal pressure on vasomotion. While recording red cell velocity in single capillaries of cat mesentery, these investigators showed that the rhythmical increases and decreases in red cell velocity caused by vasomotion of the precapillary sphincter were abolished by lowering arterial pressure. However, the vasomotor activity of the precapillary sphincter could be

II. Control of Microcirculation

reestablished during arterial occlusion by simultaneously raising venous pressure. Therefore, although blood flow was further reduced, the vasomotion of the precapillary sphincter could be restarted by increasing intraluminal pressure (Fig. 5.3).

Baez (1968) has also demonstrated myogenic activity in small blood vessels (independent of blood flow). In this study, intraluminal pressure was varied under conditions where there was no flow, and changes in arteriolar diameters were measured. It was found that some arterioles decreased their diameter when intraluminal pressure was elevated.

The mechanism of the myogenic response has not yet been fully established. It seems unlikely that the length of the smooth muscle cells is the controlled parameter in the myogenic response. Constriction of the smooth cells following an increase in transmural pressure causes a decrease in the length of the muscle cells. If the length of the smooth muscle is the controlled parameter, it is difficult to see how the vessels could maintain a constant constriction when transmural pressures increases, since the length of the muscle cell is decreased and myogenic stimulation is lost. Johnson and Wayland (1967) have proposed wall tension as being the controlled parameter in the myogenic response. According to this hypothesis, vascular smooth muscle is composed of two series-coupled functional units, i.e., a contractile element and a passively distensible sensory element. The sensory element is stimulated by an increase in wall tension, which in turn causes an increase in the activity of the contractile element. Wall tension

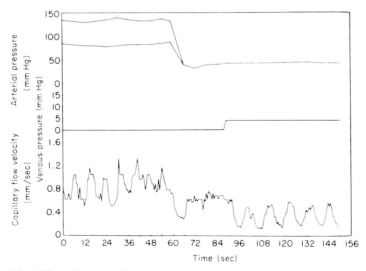

Fig. 5.3. Effect of intravascular pressure on vasomotion. A reduction in arterial pressure caused vasomotion to stop. Increasing the venous pressure restored vasomotion. (From Johnson and Wayland, 1967.)

(T) is dependent on the transmural pressure (P) and radius of the vessel (R) as described by the Law of LaPlace.

$$T = P \cdot R$$

Therefore, from the above equation, an increase in transmural pressure would increase wall tension. This increase causes stretching of the sensor element. Following contraction of the smooth muscle, with a subsequent decrease in the radius, the sensor element assumes its initial length, and the contractile element remains at its new length. The blood vessel remains constricted.

The myogenic responsiveness of the vascular smooth muscle may be important not only in controlling flow through the capillaries, but may also be of great importance in controlling the amount of fluid filtered out of the capillaries. An increase in transmural pressure would increase filtration out of the capillaries, resulting in edema formation. The myogenic responsiveness of the precapillary vessels would tend to reduce hydrostatic pressure in the capillaries and therefore attenuate edema formation.

D. Humoral Control

Humoral or blood-borne factors can also influence the activity of vascular smooth muscle as they circulate in the blood. There are a number of vasoactive substances found in cells or formed in tissues which effectively alter blood bessel diameters (Table 5.2).

TABLE 5.2

Major Humoral Substances Regulating Blood Flow in the Microcirculation[a]

Agent	Response
Catecholamines	
Epinephrine	vcs, vdil
Norepinephrine	vcs
Amines	
Serotonin	vcs, vdil, inh[b]
Histamine	vdil, inh
Polypeptides	
Angiotensin	vcs, modifier
Kinins	vdil, inh
Vasopressin	vcs, modifier
Oxytocin	vcs, vdil, modifier

[a] Modified after Altura, 1978.

[b] Thought to prevent release of norepinephrine and inhibit vasoconstrictor substances in non-relaxant doses.

II. Control of Microcirculation

It will be noted from the table that both vasoconstriction and vasodilation have been reported for a number of these materials. All of the experimental evidence for these responses cannot be presented here, but consideration should be given to differences in the responses of vessels that might occur depending on the species of experimental animal and the particular vascular bed in which the observations were made. Care must be taken to properly evaluate not only the experimental animal but the experimental design in assessing the vasoactive response. Important differences probably exist between *in vivo* and *in vitro* observations as well as between physiological and pharmacological studies. A very attractive proposal, made by Vanhoutte (1978), is that vascular smooth muscle is more homogeneous than heterogenous (although its variable responses would indicate otherwise), and that the variations result from the numerous external influences such as preload, environment, dietary factors, hormonal status, and metabolic conditions. Vanhoutte further proposes that if identical conditions could be achieved, all blood vessels would respond in an identical way.

1. *Histamine*

Histamine is contained in most cells which are in close proximity to blood vessels. Injury or damage to tissues causes histamine to be liberated, with the result that arterioles relax and capillary permeability is increased, presumably by retraction of endothelial cells. It has been suggested that the vasodilation produced by histamine is through indirect means rather than by a direct effect on vascular smooth muscle (Vanhoutte, 1978) in that histamine can inhibit the release of norepinephrine resulting in a decrease in sympathetic tone. In addition, an alteration of extracellular fluid because of increased capillary permeability may produce vasodilation.

Vasoconstriction occurs in larger arterial and venous vessels exposed to histamine. Cerebral, renal, and umbilical vessels have a stronger response to histamine than to norepinephrine. For further information about the vascular responses to histamine, see Altura and Halevy (1977).

2. *Serotonin*

5-Hydroxytryptamine (serotonin) is most often associated with vasoconstriction, although it does have a vasodilator effect in certain vascular beds. 5-Hydroxytryptamine has an important role in hemostasis, being released from platelets following vascular injury and causing constriction of smaller arteries and arterioles to reduce blood flow at the injured site. It is thought not to have any regulatory role in local control of blood flow in normal physiological conditions. The ability of this substance to cause vasodilation of small arterial vessels in skeletal muscle, mesentery, and brain may be a secondary effect.

3. Kinins

The kinins, of which bradykinin is the most often studied, are polypeptides which cause vasodilation, reduce systemic blood pressure, and may increase capillary permeability. Kinins have been shown to be very strong vasodilators when topically applied to rat microvessels. The kinins, however, have potent contractile effects on umbilical arteries and veins of pregnant women. Bradykinin evokes the release of prostaglandins from certain organs. Currently, there is not enough experimental evidence to assign a role for local regulation of blood flow to the kinins.

4. Angiotensin

Angiotensin is a polypeptide which is found in three active forms in the body (angiotensin I, II, III) of which angiotensin II, the most potent vasoactive form, causes vasoconstriction. Angiotensin II formation occurs through the following sequence.

$$\text{Angiotensinogen} \xrightarrow{\text{Renin}} \text{Angiotensin I} \xrightarrow{\text{Converting Enzyme}} \text{Angiotensin II}$$

Angiotensinogen is an α_2 globulin fraction of the blood which is converted to angiotensin I through the removal of amino acids by renin. Renin was first known to be released by the juxtaglomerular walls of the kidney. Its release is associated with a decrease in blood pressure. It is now known that renin is present in most peripheral blood vessels. "Converting enzyme" is responsible for the further removal of amino acids from angiotensin I to form angiotensin II. Converting enzyme is probably present in the endothelial cells of the systemic and pulmonary circulation as well as in the plasma. Further removal of amino acids results in the production of angiotensin III.

Since renin release is associated with a decrease in blood pressure, the potent vasoconstrictor properties of angiotensin represent a feedback mechanism to compensate for the decrease in pressure. Although angiotensin is a potent vasoconstrictor of arterial vessels, venous vessels are only weakly affected. The vasoconstrictive effects of angiotensin are strongest in the skin, splanchnic region, and kidney. The effect is much smaller on arterial vessels of the heart, skeletal muscle, and brain, and blood flow to these may actually be increased due to the elevation of blood pressure by angiotensin.

In addition to its direct action, angiotensin also potentiates the vasoconstrictor effect of sympathetic stimulation on vascular smooth muscle and can stimulate the release of prostaglandinlike substances.

II. Control of Microcirculation

5. *Vasopressin*

Vasopressin is a hormone produced in the posterior portion of the pituitary gland. Although vasopressin was discovered as a result of its vasoconstrictor properties, its primary physiological function is to control water excretion by the kidneys. The vasoconstrictor action of vasopressin is only seen when it is present in much higher concentrations than is necessary for the regulation of water excretion. There are differences in the sensitivity of vessels to vasopressin both between species and between various vascular segments. Rat mesenteric arterioles have been found to be more sensitive to vasopressin than to angiotensin (Altura, 1978), although it is generally believed that angiotensin is a more potent vasoconstrictor. It is suggested by Altura that the smaller the arterial vessel, the more responsive it is to vasopressin and that small muscular venules (40 μm) may be even more responsive than arterial vessels. Aside from its direct vasoconstrictor actions, vasopressin may also potentiate the action of catecholamines.

6. *Oxytocin*

Oxytocin is another vasoactive hormone produced by the pituitary gland. The direct effect of oxytocin on microvessels is vasoconstriction. This vasoconstrictive action is potentiated in the presence of estrogen. The microvessels of male rats are less sensitive to the constrictor action of oxytocin than those of female rats. Oxytocin may also potentiate the vasoconstrictor action of catecholamines.

7. *Catecholamines*

Under normal physiological conditions, the effects of catecholamines that are naturally released from the adrenal medulla are very small when compared to the effects of catecholamines released from sympathetic nerves innervating blood vessels. Under these conditions, circulating catecholamines are probably of greater importance in the stimulation of metabolic activity in organs such as the liver, skeletal muscle, and adipose tissue, than they are in the control of vascular diameter. Under emergency conditions, however, catecholamine release by the adrenal medulla is greatly increased and reinforces the activity of sympathetic vasoconstrictor nerves. The ratio of epinephrine to norepinephrine released by the adrenal medulla is species dependent. In man, epinephrine is primarily released.

Catecholamines have a direct effect on vascular smooth muscle through activation of adrenergic receptors on the smooth muscle cells. There are two general categories of adrenergic receptors on vascular smooth muscle; α-receptors and β-receptors. Stimulation of α-receptors causes vasoconstriction, whereas stimulation of β-receptors results in vasodilation. Norepinephrine only weakly stimulates β-receptors in blood vessels, but strongly stimulates their α-receptors and, therefore, causes constriction of systemic vessels almost exclusively. Epinephrine can stimulate either α- or β-receptors and therefore can cause either vasoconstriction or vasodilation. The type of response produced by epinephrine depends on the tissue and the concentration. Low concentrations of epinephrine cause decreased resistance to blood flow in skeletal muscle, whereas high concentrations cause increased resistance to flow. In skin, epinephrine causes only vasoconstriction.

As with other circulating vasoactive agents, the sensitivity of the vessels to catecholamines varies with species, tissue type, and vessel type. In general, the sensitivity of arterial vessels to catecholamines increases with decreasing vessel size. Precapillary sphincters have the greatest sensitivity to the vasoconstrictor actions of the catecholamines, being many times more sensitive to catecholamine stimulation than arterioles.

In general, the sensitivity of small venous vessels to catecholamines is less than that of arterial vessels. Large venous vessels seem to be more responsive to catecholamine-induced vasoconstriction than venous microvessels.

III. CONTROL OF SKELETAL MUSCLE CIRCULATION

The factors which control blood flow to skeletal muscle can be listed as neural, metabolic, and myogenic, or divided as central control through the sympathetic nervous system and local control through metabolic and myogenic influence. Nervous control involves total flow to the muscle, whereas local control involves the microcirculation.

A. Neural Control

Anatomic evidence has been presented showing adrenergic nerve terminals in all segments of the arterial vasculature of skeletal muscle, with a decrease in adrenergic terminals in the smaller arterial vessels and the arteriolar vessels. Sympathetic vasoconstrictor fibers have not been demonstrated to terminate on the vascular smooth muscle of the smallest arterioles that just precede the capillaries, although fibers can be seen in the immediate interstitial space. The density of adrenergic nerve terminals on the venous side of skeletal muscle vasculature is much lower than on the arterial side. This is consistent with the finding that maximal sympathetic stimulation causes a small reduction in the capacitance of the venous vessels in skeletal muscle at times when arterial constriction is significantly high. Maximal sympathetic vasoconstriction does not completely curtail blood flow to skeletal muscle except at very low arterial perfusion pressure; this

III. Control of Skeletal Muscle Circulation

indicates an important relationship between central and local mechanisms, which interact to control blood flow to any tissue.

Stimulation of sympathetic adrenergic nerves innervating skeletal muscle can produce a decrease in blood flow to the muscle on the order of eight- to ten-fold. Denervation of skeletal muscle vasculature results in a transient increase of blood flow of two to three times which returns toward the control level when local regulatory mechanisms take over. The increase in blood flow following denervation demonstrates the tonic sympathetic vasoconstrictor activity on the arterial or resistance vessels of skeletal muscle.

The degree of control of sympathetic vasoconstrictor fibers over the smaller arteriolar vessels and the subsequent effect on the number of perfused capillaries still remains to be determined. Three different techniques have been used in an attempt to establish the effect of sympathetic stimulation on capillary blood flow. Two of these techniques, the measurement of the capillary filtration coefficient (CFC) and the permeability-surface area product (PS), employ isolated perfused skeletal muscle. To measure CFC, the isolated muscle is either weighed or placed in a plethysmograph to measure changes in tissue volume (Cobbold *et al.*, 1963). Capillary hydrostatic pressure is elevated by increasing venous outflow pressure and the volume of fluid filtered is calculated from the change in tissue volume from control. The CFC is expressed in ml/mm Hg/min/100 g. It is a measure of the ease with which fluid leaves the capillaries and is affected by either changes in capillary permeability and/or the number of perfused capillaries. An increase in tissue volume, as determined by plethysmography or increased weight of the tissue, could indicate an increase in capillary permeability, which would promote the accumulation of interstitial fluid, or it could indicate that more capillaries were filled with blood. It could also indicate a combination of these two conditions.

The permeability-surface area (PS) produce is measured by determining the arteriovenous extraction fraction of radioactive potassium or rubidium at known blood flow rates. With blood flow and arterial concentrations of tracers kept constant, variations in the venous concentration of the tracer must be due to changes in capillary permeability and/or the number of perfused capillaries. The dimensions of PS are given in ml/min × 100 g tissue. CFC is largely independent of flow, whereas PS does vary with flow rate.

The third technique used to determine the effects of sympathetic stimulation on capillary blood flow is direct observation of the skeletal muscle vasculature during sympathetic stimulation.

Experiments utilizing CFC or PS to determine changes in vascular resistance and capillary blood flow have shown a difference in the time course of changes in these two factors during sympathetic stimulation. An increase in vascular resistance, indicating vasoconstriction of the arterial vessels, is maintained throughout the period of stimulation. Capillary perfusion initially decreases during stimulation, but after a few minutes the perfusion returns to control values or is even greater as is shown by CFC or PS measurements. This difference in the duration of the resistance response and the number of perfused capillaries is a mechanism that contributes to the efficient operation of the cardiovascular system.

Increases in vascular resistance contribute primarily to the maintenance of systemic blood pressure. Although increases in skeletal muscle vascular resistance may be contributing to the maintenance of systemic pressure for the benefit of the organism as a whole, this same response may be detrimental to the skeletal muscle supplied by these constricted vessels because of the decreased blood flow. Some of the untoward effects of increasing resistance may be attenuated if the number of perfused capillaries is maintained. Therefore, the evidence that the sympathetic nervous system has only a transient influence over capillary perfusion indicates that a local mechanism must take over the maintenance of exchange under the conditions of reduced blood flow that result from sympathetic stimulation.

Direct observation through the microscope of skeletal muscle blood vessels during sympathetic stimulation has also provided evidence that short periods of stimulation of sympathetic vasoconstrictor fibers cause a decrease in the number of perfused capillaries. Eriksson and Lisander (1972) observed that a brief stimulation of sympathetic nerves innervating cat tenuissimus muscle caused constriction of arterial vessels with diameters ranging between 200 and 15 μm and a decrease in capillary blood flow. A maximal constriction occurred in vessels of approximately 30 μm where complete closure was seen. The smallest arterial vessels invested with smooth muscle in this vascular bed are 10–12 μm (presumably terminal arterioles), and they were not affected. Baez et al. (1977), using direct microscopic observation, reported that stimulation of certain areas of the brain caused a decrease in the diameter of all precapillary microvessels in the rat cremaster muscle and a reduction in the number of perfused capillaries. The duration of constriction of the vessels outlasted the duration of central stimulation. These observations suggest that liberation of neurohormonal substances produced the vasoconstriction. Eriksson and Lisander (1972), observed that, in cat tenuissimus muscle, prolonged stimulation of sympathetic nerves caused arterial vessels to constrict and relax alternately with a regular periodicity.

All three methods, the use of capillary filtration coefficient, the determination of the permeability surface area product, and direct microscopic observation, demonstrated a decreased blood flow to skeletal muscle during sympathetic stimulation with a concurrent decrease in the number of perfused capillaries initially. The influence of the sympathetic vasoconstrictor nerves over the number of perfused capillaries is, however, of little importance when the effect is only transient and is quickly superceded by local control.

III. Control of Skeletal Muscle Circulation

One outstanding point of controversy in neural control of any of the microvascular beds is the extent of nerve distribution along the arterial tree. The lack of uniformity in the location or definition of the precapillary sphincter is responsible for much of the conflict. The term "precapillary sphincter" was originally used to describe a muscular investment at the origin of capillary vessels from their parent arteriole that can control the flow of blood through the distally located capillaries. Investigators who have used whole-organ techniques to measure capillary surface area and vascular resistance in skeletal muscle adopted a less discrete usage of the term so that it signified any area of precapillary resistance. In the whole-organ studies, a calculated increase in capillary suface area without a proportional decrease in arterial resistance was interpreted to be the result of relaxation of the precapillary sphincter. A decrease in capillary surface area without a proportional increase in arterial resistance was interpreted as a constriction of the precapillary sphincter. Since these investigators were unable to observe changes in the diameter of vessels in specific segments of the vascular tree, the precapillary sphincter represented a functional rather than an anatomical term.

Investigators using direct observation of skeletal muscle blood flow through the microscope reported that there were often rhythmical variations in capillary blood flow. It was also found that these rhythmical variations in flow occurred simultaneously in large groups of capillaries, often more than ten vessels in a group (Fig. 5.4).

Flow variations were never restricted to single or even pairs of capillaries. In referring back to the original definition of the precapillary sphincter, if perfusion of the capillaries was controlled at the origin of each capillary, one would not expect large numbers of capillaries to have periodic flow in synchrony. Since

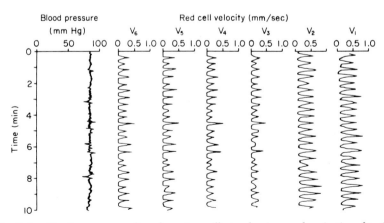

Fig. 5.4. Simultaneous recordings from six capillaries showing synchronization of periodic flow. (From Tuma *et al.*, 1975.)

there was no evidence for control of perfusion of a single capillary, it was concluded that no precapillary sphincter was located at the origin of individual capillaries in this bed.

In order to resolve some of the controversy arising from a nonuniform usage of the term precapillary sphincter, and to make the interpretation of results using various techniques more understandable, we have previously advocated (Wiedeman *et al.*, 1976) and in this book have adopted a more restricted definition for the precapillary sphincter. The precapillary sphincter refers to the last smooth muscle cell in the arterial vasculature preceeding the capillary. We have adopted this usage since this cell represents the final potential control point for the determination of the number of perfused capillaries. Also, the opening and closing of the most distal arterial vessels would normally contribute little to total changes in arterial resistance since the greatest pressure drop occurs upstream from this location. In adopting this definition for the precapillary sphincter, one must stress that this last potential site of control of the number of perfused capillaries is governed by local control.

Thus far, we have dealt with control of skeletal muscle vascular resistance and blood flow by sympathetic adrenergic nerves. In a number of species, sympathetic cholinergic nerves also exert an influence over skeletal muscle blood flow and vascular resistance. Morphologic and functional studies have demonstrated the presence of sympathetic cholinergic fibers innervating arterial blood vessels in the skeletal muscle of the cat, dog, and sheep. These fibers have not been shown to exist in the rabbit, rat, monkey, or man. In the species where they are present, sympathetic cholinergic innervation of skeletal muscle blood vessels is probably restricted to the arterial vessels. Stimulation of these fibers produces a large increase in skeletal muscle blood flow, lasting for approximately 20 seconds. Following this, there is a smaller persistent decrease in resistance lasting for at least 2 minutes. Although stimulation of these fibers causes a brief, pronounced increase in skeletal muscle blood flow, the number of capillaries perfused with blood is not increased. Therefore, sympathetic cholinergic stimulation does not result in an increase in the diffusion of material across the exchange vessels. The reason that the capillary surface area available for exchange does not increase with sympathetic cholinergic stimulation probably arises from the fact that these fibers only innervate the large arterial vessels. When these large vessels are dilated, smaller arterial vessels downstream tend to constrict because of myogenic stimulation and a decrease in metabolic vasodilators. It has been postulated that sympathetic cholinergic fibers are involved in the preparation for exercise and defense reactions.

B. Myogenic Control

Blood vessels in skeletal muscle respond directly to changes in transmural pressure, constricting when transmural pressure increases and dilating when

III. Control of Skeletal Muscle Circulation

transmural pressure decreases. This myogenic autoregulation is probably most effective as an immediate response to changes in perfusion pressure and is later coupled with metabolic regulatory changes in blood flow.

C. Metabolic Control

The metabolic activity of skeletal muscle cells produces a number of vasodilator metabolites which diffuse out into the interstitial fluid surrounding the blood vessels. If blood flow to the skeletal muscle is decreased, vasodilator metabolites accumulate, resulting in vasodilation of resistance vessels to promote adequate distribution of the available blood to the muscle. Metabolic regulation is most important during increased activity of skeletal muscle. The autonomic nervous system has a large influence over blood flow to resting skeletal muscle, but metabolic control is the primary mechanism regulating skeletal muscle blood flow during increased contractile activity during exercise. A number of investigations have been made to determine which of the metabolites are primarily responsible for the increase in skeletal muscle blood flow during exercise. At the present time no single factor has been found which could alone account for the increase. It appears that the combined action of a number of metabolites are responsible for the increase. The interstitial fluid concentration of potassium, CO_2, H^+, phosphate, adenosine, adenine nucleotides, and magnesium are increased during contraction of skeletal muscle. All of these substances are vasodilators and may contribute to the active hyperemia. None of these substances alone, when present in the concentration produced physiologically, is able to produce the same degree of vasodilation as that seen in active hyperemia.

D. Humoral Control

Histamine, prostaglandins, and angiotensin have all been shown to have an effect on skeletal muscle blood flow, but no regulatory role can be assigned to them.

Epinephrine is the most studied, but confirmation of the mechanisms responsible for its dual effect is still lacking. Low concentrations reduce resistance to flow in skeletal muscle, whereas high concentrations increase resistance. The decreased resistance produced by low concentrations can be abolished by the administration of a β-blocking drug, suggesting that β-receptors will react at a concentration of epinephrine too weak to stimulate α-receptors. High concentrations of epinephrine cause an increase in peripheral resistance to skeletal muscle blood flow through stimulation of α-receptors.

E. Capillary Perfusion in Skeletal Muscle

An important aspect of skeletal muscle blood flow is the distribution of the total flow to the numerous capillaries that run parallel to the muscle fibers.

118   5. Factors Involved in the Regulation of Blood Flow

Krogh, in 1922, proposed the idea that blood flow in skeletal muscle would be most efficient if only a fraction of the capillaries were perfused at rest when the demand for oxygen delivery is low. He estimated that during muscular contraction the number of capillaries perfused with blood increases eight times (Krogh, 1929). More recent studies have estimated that between one-third and one-half of

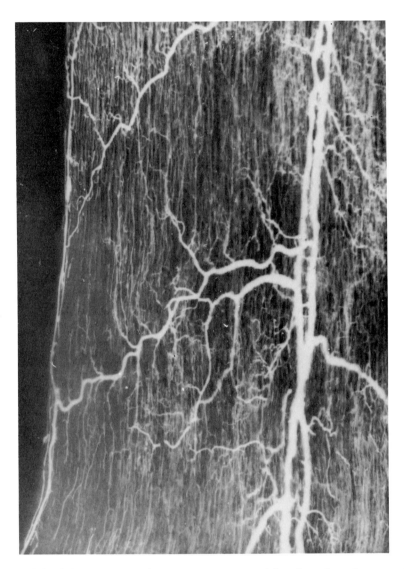

Fig. 5.5. A direct connection between a transverse arteriole and venule can be seen at the free edge of the tenuissimus muscle.

the capillaries are perfused during resting conditions (Lindbom *et al.* 1980). Although there is no agreement currently as to the exact fraction of capillaries perfused at rest, most investigators would agree that there is recruitment of capillaries in exercising muscle.

It has also been shown that under certain circumstances, the amount of exchange occurring across the capillaries in skeletal muscle is not proportional to skeletal muscle blood flow (Barcroft and Dornhorst, 1954; Walder, 1953, 1955, 1968; Barlow *et al.*, 1958, 1959, 1961). The existence of a nonnutritional or shunt pathway for blood flow has been proposed as an explanation for these observations. The nutritional pathway affords the passage of blood through capillaries in which an exchange of materials occurs as blood flows through them. Nonnutritional channels could be composed of large arteriovenous anastomoses, short capillaries, or vessels of slightly larger diameter than the capillaries, where blood could pass from the arterial to the venous system quickly and without equilibrating with the solvents in the interstitium. Presently, the evidence indicates that there is not a significant number of large diameter arteriovenous anastomoses which could account for the nonnutritional component of skeletal muscle blood flow. A number of studies made by direct observation of flat skeletal muscle report the existence of connections between arteries and veins which are slightly larger than capillaries and have more rapid flow. These vessels are located near the free edge of the muscle (Fig. 5.5) or in the adjacent connective tissue.

In the spinotrapezius and tenuissimus muscle, flow is seen to continue in these vessels under low flow conditions when flow has ceased in the capillaries (Zweifach and Metz, 1955).

IV. CONTROL OF CEREBRAL CIRCULATION

The control of cerebral blood flow differs to some degree from control of flow to other organs, perhaps because of the necessity for maintaining a relatively large and stable flow to this highly metabolically active tissue. Deprivation of blood flow is less well tolerated by the central nervous tissue than any other tissue, and the cardiovascular system is arranged to ensure adequate delivery of blood.

A. Neural Control

Blood vessels of the brain are innervated by both sympathetic and parasympathetic nerves. The importance of these nerves in controlling cerebral blood flow, however, appears to be minimal under normal physiological conditions. Extensive reviews have been presented by Heistad and Marcus (1978) and Purves (1978).

The density of sympathetic innervation is greatest in the larger blood vessels of the brain, although arterioles as small as 15 μm in diameter are reported to be innervated (Edvinsson, 1975). It has been postulated that the function of the nerves innervating the large cerebral vessels may be to regulate the cerebral vascular volume and, therefore, the intracranial pressure. The sensitivity of pial arterial vessels to the vasoconstrictor actions of norepinephrine is low in comparison to the sensitivity of arterial vessels in other organs of the body. Kontos (1975) reports that pial arterial vessels smaller than 100 μm in diameter are unresponsive to supramaximal sympathetic stimulation and to topically applied norepinephrine.

Innervation of cerebral blood vessels by parasympathetic nerves is restricted to the surface of the brain and does not extend into the parenchyma of the cortex where the greatest resistance occurs. Although the cerebral vessels dilate in response to acetylcholine and constrict in response to norepinephrine, vascular diameter remains unchanged when the receptors for these neurotransmitters are blocked, indicating that the autonomic nervous system does not exert a tonic effect on cerebral vascular diameter.

The exact function of autonomic nerves that terminate on cerebral blood vessels remains unclear. It seems unlikely that the regulation of cerebral blood flow by these nerves is of great importance under normal physiological conditions. The autonomic nerves of cerebral blood vessels may, however, play an important role in protecting the brain under pathological conditions such as hypertension.

B. Metabolic Control

Cerebral blood flow is closely regulated by the metabolic activity of the brain. Although total cerebral blood flow remains relatively constant under normal physiological conditions, it has been demonstrated that localized changes in cortical activity cause a redistribution of blood flow within the cerebral cortex.

The resistance vessels of the brain are extremely sensitive to changes in carbon dioxide tension. Increasing P_{CO_2} in either the blood or the cerebral spinal fluid causes a pronounced vasodilation and an increase in cerebral blood flow, most likely mediated through changes in H^+ concentration (Kontos et al., 1971) (Fig. 5.6).

The close association that exists between arterial P_{O_2} and blood flow in skeletal muscle or cardiac muscle is not seen in the cerebral circulation. In the normal physiological range, changes in arterial P_{O_2} do not cause significant changes in cerebral blood flow. Cerebral blood flow does not significantly increase until arterial P_{O_2} falls below 50 mm Hg. The inhalation of 80–90% oxygen causes only a small (12%) decrease in cerebral blood flow (Kennedy et al., 1971–1972) (Fig. 5.7).

IV. Control of Cerebral Circulation

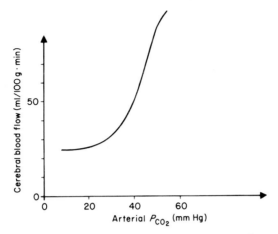

Fig. 5.6. Variations in cerebral blood flow as a function of acute changes in arterial P_{CO_2}. (From Lassen, 1978.)

The changes in cerebral blood flow caused by hypoxia and hypercapnia are both mediated through changes in hydrogen ion concentration.

Since the concentration of K^+ and adenosine in the cerebral spinal fluid increases with decreasing cerebral blood flow or increased cerebral activity, it has been suggested that these substances may be involved in the local regulation of cerebral blood flow. The experimental evidence available does not permit an evaluation of the physiologic influence of these substances.

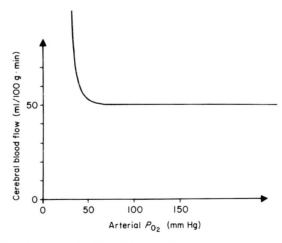

Fig. 5.7. Variations in cerebral blood flow as a function of changes in arterial P_{O_2}. (From Lassen, 1978.)

V. CONTROL OF CARDIAC BLOOD FLOW

As is the case with the brain, the survival of the body is dependent on maintaining blood flow to the heart at a level adequate to meet the metabolic needs of this continually active tissue. It is not surprising, therefore, that those factors which tend to ensure perfusion of the coronary microcirculation predominate in the control of myocardial blood flow.

A. Neural Control

The direct influence of the autonomic nerves on coronary blood vessels is relatively unimportant under normal physiological conditions in comparison to the other factors involved in the regulation of coronary blood flow. The direct effect of stimulation of the sympathetic nerves innervating the heart is vasoconstriction. This vasoconstriction is overridden by the influence of the increased metabolic activity of the heart, which also results from sympathetic stimulation. The direct effect of sympathetic nerves on the coronary blood vessels can be seen experimentally by eliminating the metabolic effect of sympathetic stimulation of the heart. This can be done by selectively blocking the adrenergic receptors of the cardiac muscle, which differ from the adrenergic receptors of vascular smooth muscle. Under this condition, the coronary vascular resistance increases during sympathetic stimulation.

B. Metabolic Control

Metabolic control is the most important mechanism regulating myocardial blood flow. Coronary blood flow is directly related to tissue oxygen tension. Although myocardial hypoxia is the most powerful stimulus for coronary vasodilation, it is unlikely, as discussed previously, that oxygen directly regulates blood vessel diameter in the heart. An increase in the metabolic activity of the heart results in elevation of P_{CO_2} and lactic acid concentration and a decrease in pH in the interstitial fluid of the myocardium. Although all of these factors can cause some degree of dilation of the coronary vasculature, they cannot, in the concentrations produced physiologically, account for the large increase in coronary blood flow that occurs during increased metabolic activity.

At the present time, adenosine seems to be the most important factor in coupling myocardial metabolism with myocardial blood flow. Adenosine can, in the concentrations achieved in the normally functioning heart, account for the magnitude of hyperemia which occurs during increased metabolic activity of the heart or during myocardial hypoxia (Rubio and Berne, 1978).

C. Myogenic Control

At the present time there is not enough information available to evaluate the degree to which myogenic control may participate in the regulation of coronary blood flow.

D. Perfusion of the Capillaries in the Heart

The density of capillaries in the heart is very high in comparison to that in skeletal muscle. This gives the microcirculation of the heart a more efficient exchange of materials across the capillaries and accounts for the high arteriovenous oxygen difference in the coronary circulation. Studies by Martini and Honig (1969) indicate that approximately half of the capillaries in the unstressed heart are perfused with blood. When the metabolic activity of the heart is increased or when the oxygen tension of the blood perfusing the heart is reduced, more capillaries become perfused with blood and there is an increased efficiency of exchange. The heart, therefore, has two mechanisms to increase oxygen delivery; one to decrease resistance to blood flow, and the other to increase the number of capillaries perfused with blood.

Small arterio-arterial anastomoses exist between arterial vessels in the heart and offer a collateral channel for blood flow, which helps to some degree to ensure perfusion of the capillaries. These anastomoses are relatively small vessels—less than 40 μm in the canine heart (Schaper, 1971), and therefore cannot compensate for acute occlusion of large coronary arterial vessels. When occlusion of large coronary arterial vessels occurs slowly, over months and years, the collateral vessels increase in diameter and can compensate to a greater degree for the occlusion.

VI. CONTROL OF GASTROINTESTINAL CIRCULATION

When considering the control of blood flow to the alimentary canal, one must keep in mind that the microcirculation in this area has important functions in addition to meeting the metabolic needs of the tissue. This is in contrast to tissues such as the heart and brain where the function of the microcirculation is almost exclusively to meet the metabolic requirements of these tissues. The gastrointestinal microcirculation must provide solute for the secretory processes of the gut, and it is involved in the absorption of materials from the gut lumen.

As illustrated in Fig. 5.8, the vasculature of the gut wall can be thought of as being comprised of three parallel circuits. These parallel circuits have the capa-

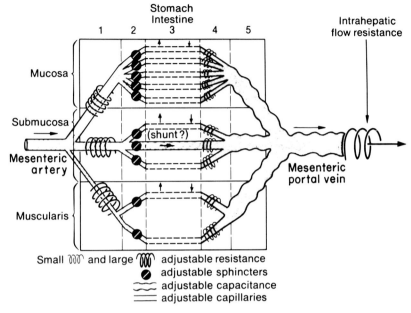

Fig. 5.8. A schematic illustration of the vasculature of the gastrointestinal tract. The vessels supplying the mucosa, submucosa, and muscularis are arranged in parallel creating the possibility for separate control of flow to each of the three compartments. (From Folkow, 1967.)

bility to independently regulate blood flow to each of the major layers of the alimentary canal, according to their functional needs.

The distribution of blood flow to each of the major layers is heterogeneous, as is illustrated in Fig. 5.9. This heterogeneity of flow is a reflection of the variations in the functional needs of the tissues comprising the wall of the alimentary canal.

A. Neural Control

Stimulation of sympathetic nerves innervating the stomach and intestines causes an increase in vascular resistance and a decrease in venous capacitance. Bohlen and Gore (1979) report a decrease in the diameter and the intraluminal pressure of all arterial and venous vessels in the intestinal muscle microcirculation during sympathetic stimulation. Although there is a significant increase in gastrointestinal vascular resistance following sympathetic stimulation, this response is only transient; flow nearly returns to control values after 2–4 minutes of stimulation. This attenuation of the influence of the sympathetic nerves on the resistance vessels during sustained stimulation is called "autoregulatory es-

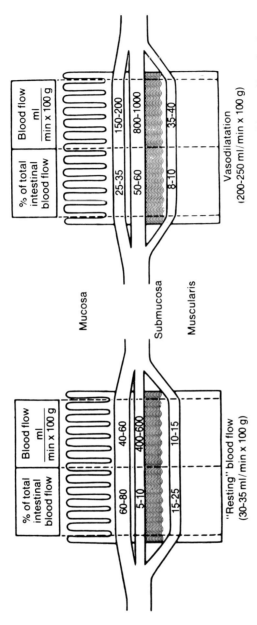

Fig. 5.9. Changes in the distribution of blood flow in the small intestine at rest and following vasodilation. (From Lundgren, 1967.)

cape." The term autoregulatory may be somewhat misleading, since the mechanisms normally associated with autoregulation, i.e., myogenic and metabolic control, cannot totally explain this response. The accumulation of vasodilator metabolites is not thought to be the primary cause for autoregulatory escape, since this phenomenon also occurs in tissues perfused with constant flow during sympathetic stimulation. Measurements of CFC and PS during sympathetic stimulation indicates that a reduction in the number of capillaries perfused with blood persists even after autoregulatory escape has occurred (Dresel et al., 1966).

Unlike the response of the resistance vessels, autoregulatory escape of the capacitance vessels does not occur during continuous stimulation. Since the increase in resistance is only of short duration, whereas the capacitance change is maintained, the change in capacitance of the gastrointestinal tract is the most important contribution by this area to the regulation of systemic pressure. It is estimated that almost one-half liter of blood may be mobilized from the gastrointestinal tract by the decrease in capacitance of the vessels that occurs during sympathetic stimulation.

Stimulation of parasympathetic fibers innervating the stomach and colon result in an increase in blood flow. This increase in flow, however, is not due to a direct effect of the parasympathetic nerves on the blood vessels. Parasympathetic nerve stimulation causes an increase in gastrointestinal secretion, which secondarily causes an increase in blood flow.

B. Metabolic and Humoral Control

In addition to supplying nutrient materials for the parenchymal tissue, the gastrointestinal circulation plays an important role in the absorptive and secretory functions of the gut. Gastrointestinal blood flow closely parallels secretory activity. The increase in blood flow that occurs when secretory activity is increased is probably the result of a combination of the production of vasoactive hormones and vasodilator metabolites. The vasoactive hormones may either directly cause vasodilation or may cause relaxation of vascular smooth muscle secondarily through increased metabolism. Gastrin, secretin, and cholecystokinin all increase mucosal blood flow. The increase in blood flow in response to gastrin is probably due to an increase in tissue metabolism, whereas secretin and cholecystokinin cause vasodilation directly and through changes in metabolism (Lundgren, 1978). The responsiveness of the intestinal vasculature to changes in metabolism is demonstrated by the sensitivity of these vessels to hypoxia (Mohamed and Bean 1951; Korner et al., 1967).

The influence of changes in gastrointestinal motility on blood flow has not been clearly established. Increasing gastrointestinal motility could have two opposite effects on blood flow. Contractions of the gut may compress the blood

VI. Control of Gastrointestinal Circulation

vessels, which tends to restrict blood flow, whereas the accumulation of vasodilator metabolites tends to increase flow. More precise measurements of local blood flow in various layers of the gut wall during motility changes are required before the relationship between motility and blood flow can be clearly established. It is known, however, that intestinal smooth muscle has less potential than skeletal muscle to increase blood flow above resting levels. Upon maximal vasodilation, blood flow to resting gastric smooth muscle only increases by a factor of about six, whereas skeletal muscle blood flow may increase over twenty times.

Vasopressin and epinephrine both cause a reduction in gastrointestinal blood flow. Their effect on tissue oxygen consumption, however, is not similar, indicating a difference in their influence on blood flow distribution. Vasopressin decreases oxygen consumption, whereas epinephrine can cause blood flow to decrease without affecting oxygen consumption (Pawlik *et al.*, 1975).

C. Myogenic Control

Myogenic control is thought to play a significant role in the local regulation of gastrointestinal blood flow (Johnson, 1971; Lundgren, 1978). Precise regulation of capillary hydrostatic pressure is especially important in the mucosal layer. Capillaries of the mucosa are fenestrated and have a high hydraulic conductivity. Because of their high hydraulic conductivity, an increase in mucosal capillary hydrostatic pressure could lead to edema formation and a large loss of volume from the vascular system.

Although vascular pressure in the mucosa is thought to be precisely regulated, this does not seem to be the case for vessels in the intestinal smooth muscle. Gore and Bohlen (1975) found that pressure in arterioles, capillaries, and venules all decreased in direct linear proportion to the reduction in perfusion pressure. This would indicate poor myogenic control of pressure in this area.

D. Countercurrent Mechanisms of the Villus

The anatomical arrangement of the villus arteriole and capillary network provides an opportunity for countercurrent exchange in the intestine, as illustrated in Fig. 5.10. Lipid soluble substances, water, and low molecular weight ionic solutes would be most affected by this countercurrent exchange mechanism (Lundgren, 1978).

The principles governing countercurrent exchange in this area would be the same as those classically described in the kidney. Materials absorbed at the tip of the villus are carried back to the venous blood by the villus capillary network. Since there is only a 20 μm separation between the villus capillaries and the arteriole in the center of the villus, those absorbed substances which can easily

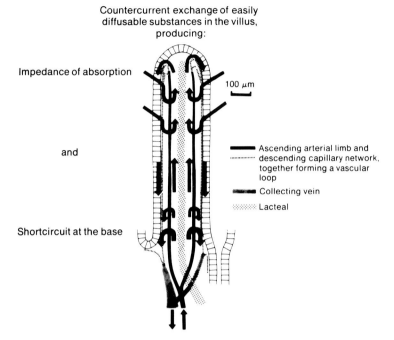

Fig. 5.10. A schematic diagram of the anatomical arrangement of the microvessels in the intestinal villus. The close approximation of the ascending arteriole and descending capillary network creates the possibility for a countercurrent exchange process. (From Lundgren, 1967.)

pass across the capillary endothelium, such as long chain fatty acids and low molecular weight solutes, will move down their concentration gradients into the lumen of the arteriole. This would cause part of the absorbed material to be returned to the tip of the villus and would result in an impedance of absorption of material from the intestinal lumen. Substances absorbed by the villus would be partially trapped at the tip, creating a concentration gradient with the highest concentration near the tip of the villus. This is thought to be the case with absorbed sodium (Lundgren, 1978). The concentration gradient of solutes such as sodium would, of course, also result in an osmotic gradient along the villus. The high osmolarity of the villus tip may be the most important consequence of the countercurrent exchange mechanism, since this would aid in the absorption of water from the intestinal lumen.

The villus countercurrent exchange mechanism may work in the opposite direction for substances already present in the arterial blood supplying the intestine, such as oxygen. Oxygen, which easily diffuses out of the arterioles, would be taken up by the capillary blood returning from the tip of the villus, causing a "shunting" of oxygen away from the villus tip. This would result in the villus tip being slightly hypoxic even under normal conditions and may account for the rapid turnover of epithelial cells in this area. Under conditions of low flow, the time for countercurrent exchange would be even longer, with the result that an even greater shunting of oxygen away from the villus tip would occur.

VII. CONTROL OF CUTANEOUS CIRCULATION

The regulation of the cutaneous circulation, in contrast to the regulation of cardiac and cerebral blood flow, is not primarily directed at matching the metabolic needs of the tissue. The principal function of the cutaneous circulation is the regulation of body temperature. During exposure to warm ambient temperatures, blood flow to the skin exceeds that necessary to sustain the metabolic needs of the tissue, but is necessary to aid in heat loss.

The presence of a large number of arteriovenous anastomoses in certain areas of the skin also creates important differences between the regulation of flow through the cutaneous microcirculation and the regulation of microvascular flow in other areas. Skin arteriovenous anastomoses are located in the fingers, palms, feet, ears, nose, and lips in humans. When open, these vessels offer a low resistance pathway for blood flow which bypasses the capillary exchange vessels.

A. Neural Control

The temperature-regulating centers of the hypothalamus exert direct control over cutaneous blood flow through sympathetic vasoconstrictor fibers innervating arterial resistance vessels, venous vessels, and arteriovenous anastomoses; indirect control is exerted through sympathetic fibers innervating sweat glands in the skin. Alterations in sympathetic activity can cause hand blood flow to vary from nearly zero to over 75 ml/100 g/min.

The diameter of the arteriovenous (a-v) anastomoses is almost entirely determined by the activity of the sympathetic vasoconstrictor fibers innervating these vessels. An inhibition of the activity of sympathetic nerves innervating the a-v anastomoses results in vasodilation and a large increase in blood flow in apical areas. Opening of these a-v anastomoses increases the volume of blood filling the venous plexus of the skin, which aids in the conduction of heat to the body surface. Stimulation of the sympathetic nerves innervating the a-v anastomoses causes these vessels to close, decreasing the filling of the venous plexus and the heat loss at the body surface.

Although the arterial resistance vessels are more responsive to local influences than are the a-v anastomoses, their diameters are also primarily controlled by the activity of the sympathetic nerves. A decrease in body temperature results in an activation of the sympathetic fibers innervating these vessels, causing them to constrict. The volume of blood contained in skin venous vessels is also influenced by sympathetic activity; stimulation causes pronounced decreases in skin blood volume.

A second pathway for regulation of cutaneous blood flow by the sympathetic nervous system is through changes in the activity of sympathetic cholinergic fibers innervating the sweat glands. It has been proposed that stimulation of these fibers causes the release of kallikrein which results in the formation of bradykinin. Bradykinin, a potent vasodilator, may then cause dilation of arterioles, resulting in an increase in skin blood flow (Fox and Hilton, 1958). It has also been proposed that prostaglandins may be involved in this vasodilator pathway (Heistad and Abboud, 1974).

It has been suggested that an axonal reflex, as illustrated in Fig. 5.11, may be involved in the local regulation of cutaneous blood flow. The nerves involved in this reflex are unmyelinated nociceptor fibers which have collateral branches to nearby arterioles. Stimulation of these receptors is thought to cause the release of an unknown vasodilator substance which produces a long-lasting vasodilation of the arterioles. Among the substances proposed as possible neurotransmitters in this reflex are bradykinin, histamine, and ATP.

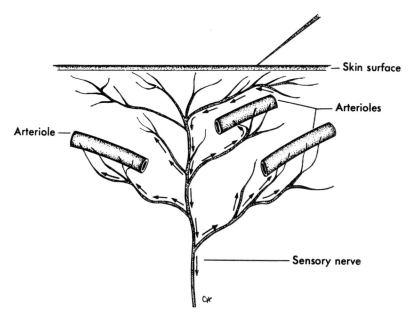

Fig. 5.11. A schematic representation of the axonal reflex involving nociceptive fibers in the skin. (From Berne and Levy, 1977.)

B. Local Control

The direct effect of changes in temperature on skin blood vessels is one of the most important local factors involved in the regulation of cutaneous blood flow. Cooling of skin arterial and venous vessels causes vasoconstriction, whereas local warming results in relaxation of vascular smooth muscle. Although the neuronal reflexes involved in the control of body temperature are more potent, this local effect of temperature can reinforce the reflex-induced changes in vascular diameter. Prolonged exposure (2–10 minutes) to temperatures less than 15°C results in a phenomenon called "cold vasodilation." In this case, there are periods of intense vasoconstriction alternating with periods of vasodilation. The dilation is thought to be the result of a direct inhibition of the contractile mechanisms of the vascular smooth muscle (Sparks, 1978).

Since the metabolic rate of skin is very low, it is felt that under most conditions, the accumulation of vasodilator metabolites is not an important mechanism of local control. Transient occlusion of blood flow to the skin does result in a reactive hyperemia which is thought to be primarily the result of a myogenic reflex (Berne and Levy, 1977). The fact that cutaneous vascular resistance returns toward control following sympathectomy also supports the concept that skin blood vessel diameter is influenced by the myogenic properties of the vascular smooth muscle.

VIII. AUTOREGULATION

The combined effects of myogenic and metabolic control mechanisms are responsible for the phenomenon of autoregulation, which is seen to occur to some degree in most organs of the body. The term autoregulation has been defined by Johnson (1974) as the maintenance of a constant blood flow in the face of changes in perfusion pressure. Although the sympathetic nervous system may modulate the range of pressures over which an organ may autoregulate, the presence of the sympathetic nerves is not necessary for this phenomenon to occur. The relative contribution of myogenic and metabolic control mechanisms to autoregulation may vary between organs. The magnitude of autoregulation also varies greatly from organ to organ.

The ability of the skeletal muscle vasculature to autoregulate is illustrated in Fig. 5.12, which shows an experiment in which the velocity of blood in the capillaries of the tenuissimus muscle was measured during occlusion of the distal aorta. It can be seen from these recordings that capillary blood flow remained constant when perfusion pressure was as low as 20 mm Hg. The reactive hyperemia that occurred following sudden removal of occlusion is a result of the dilation of the arterioles, which occurred when perfusion pressure was lowered.

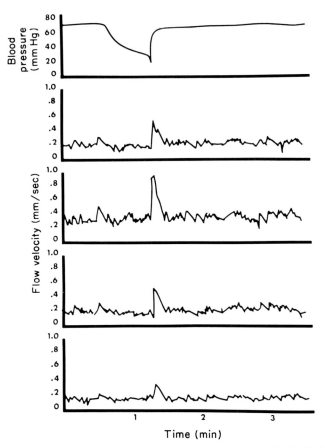

Fig. 5.12. Simultaneous tracings of hindlimb perfusion pressure and red blood cell velocity in four capillaries in the cat tenuissimus muscle. Although perfusion pressure was decreased by partial occlusion of the aorta, capillary RBC velocity did not decrease. Following the removal of occlusion a reactive hyperemia occurred. (From Tuma et al., 1977.)

Figure 5.13 shows that isolated perfused skeletal muscle shows good autoregulatory capability when perfusion pressure ranges between 20 and 160 mm Hg. Both metabolic and myogenic control mechanisms seem to contribute to autoregulation in skeletal muscle (Tuma et al., 1977).

In normal individuals, cerebral blood flow remains relatively constant over a mean arterial pressure range of 60–130 mm Hg. The autoregulation of cerebral blood flow is probably due to a combination of metabolic and myogenic control mechanisms. Kontos (1975) made a study of the participation of pial arterial vessels in the autoregulatory response. It was determined that large pial arterial vessels (> 100 μm) were involved in the autoregulatory response when mean

VIII. Autoregulation

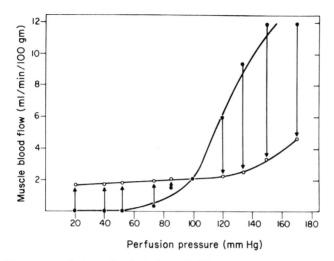

Fig. 5.13. Autoregulation in dog skeletal muscle. The closed circles represent the transient alterations in flow following an abrupt change in perfusion pressures. The open circles represent the steady state flow at the new perfusion pressure. (From Berne and Levy, 1977.)

arterial pressure varied between 80 and 100 mm Hg. Smaller pial arterial vessels (30–60 μm) only became involved in the autoregulation of blood flow when mean arterial pressure decreased below 80 mm Hg. These vessels continue dilating until mean arterial pressure becomes lower than 45 mm Hg.

During chronic hypertension, the autoregulatory curve becomes shifted to the right, with cerebral blood flow being maintained at a normal value, even when the cerebral vasculature is exposed to pressures that would cause an increase in cerebral blood flow in normal individuals. This shift in the pressure range over which autoregulation is possible has an obvious protective effect for the brain during hypertension. However, along with this greater tolerance of high pressures, there is a decrease in the ability of the brain to maintain flow when mean arterial pressure is lowered (Fig. 5.14).

As in other organs, the autoregulation of cerebral blood flow occurs even in the absence of nerves.

Experimental evidence either demonstrating or refuting the presence of a myogenic response of cerebral blood vessels *in vivo* is lacking. A definitive statement concerning the importance of myogenic control is not possible, since experiments made thus far have not permitted a separation of myogenic and metabolic control mechanisms. The rapid changes in cerebral vascular resistance following a decrease in arterial pressure gives support to the idea that myogenic control may combine with metabolic control in the regulation of cerebral blood flow.

The coronary vascular bed has good autoregulatory control between perfusion

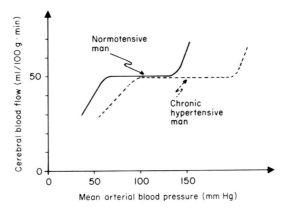

Fig. 5.14. The shift in the cerebral autoregulatory curve of blood flow that occurs in hypertensive individuals. (From Lassen, 1978.)

pressures of 60 and 190 mm Hg. Experimental evidence, provided by Berne and Levy (1977) is presented in Fig. 5.15.

The ability of the coronary vessels to autoregulate does not appear, however, to be uniform in all areas of the beating heart. This is the result of the fact that vessels near the endocardial surface are compressed more by the contraction of the ventricles than vessels near the epicardial surface. Because of the larger extravascular resistance to blood flow near the endocardium, vessels in this area must be more dilated normally than vessels near the epicardium in order to deliver adequate blood flow. It has been demonstrated that during systole, flow to the endocardium is lower than flow to the epicardium. During diastole this situation is reversed, with the endocardium having a greater flow (Downey et al., 1974). The net result is that over an entire cardiac cycle, flow throughout the ventricular wall is uniform. Because, under normal conditions, the endocardial vessels are already partially dilated, the vessels in this area cannot compensate for a reduction in perfusion pressure as well as vessels located near the epicardium. Therefore, epicardial vessels are able to autoregulate at lower perfusion pressures. The result is that when perfusion pressure is reduced, e.g., during hypotension, the endocardial portion of the ventricle is more susceptible to ischemia.

The autoregulatory capability of the gastrointestinal circulation is not as pronounced as that of the vasculature in other areas, such as brain and skeletal muscle. There is a 90% reduction in intestinal blood flow when perfusion pressure is lowered from 120 to 160 mm Hg (Johnson, 1960). In both skeletal muscle and the brain there is almost no change in flow over this pressure range. The autoregulatory ability of the large intestine seems to be even less than that of the small intestine. The autoregulatory capability of the intestine does not seem to be uniform in all areas of the intestinal wall. Lundgren and Sranik (1973) found

IX. Summary

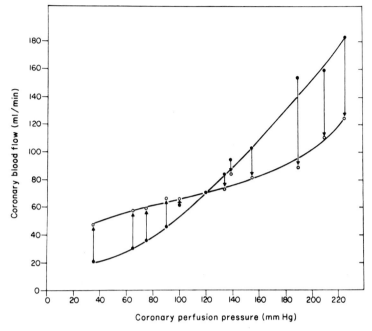

Fig. 5.15. Pressure–flow relationships in the coronary vascular bed. (From Berne and Levy, 1977.)

little change in blood flow in the villus vessels over a perfusion pressure range of 100–30 mm Hg. This indicates that the autoregulatory capability of the mucosa is much greater than that of other intestinal wall layers.

The autoregulatory capability of the cutaneous vasculature is much more limited than that of other organs such as skeletal muscle, myocardium, and brain. Autoregulation does not occur when perfusion pressure is lower than 100 mm Hg (Collins and Ludbrook, 1967). Autoregulation of blood flow in the hand occurs to some degree between perfusion pressures of 100 and 120 mm Hg and in the toes between 100 and 170 mm Hg (Shepherd, 1963). Since the accumulation of vasodilator metabolites is thought to have little influence over cutaneous blood flow, myogenic control is felt to be the principal mechanism responsible for cutaneous autoregulation.

IX. SUMMARY

Regulation of blood flow at the microcirculatory level occurs through alterations in vascular diameter. The factors which influence vascular diameter can be divided into four general categories; neural, humoral, metabolic, and myogenic. The relative importance of each of these factors in controlling vascular diameter varies among the different organs of the body and may be altered by changes in the functional activity of the various tissues.

Neural control is effected primarily through changes in the activity of the sympathetic vasoconstrictor nerves. The primary function of the nervous system in this case is to maintain systemic blood pressure. Stimulation of the sympathetic vasoconstrictor nerves does not affect all segments of the microvasculature equally. Large arterial vessels are influenced to a greater degree by sympathetic nerve stimulation than are the terminal arterioles and precapillary sphincters which are under local control.

The responsiveness of the vascular smooth muscle to metabolites produced by the cells of the parenchymal tissue surrounding the blood vessels provides for a coupling between the needs of the tissue and the amount of blood delivered. Included among the substances which may be involved in the metabolic regulation of blood flow are adenosine and adenine nucleotides, oxygen, hydrogen ion, potassium, inorganic phosphate, carbon dioxide, Kreb's cycle intermediates, and possibly prostaglandins, thromboxanes and prostacyclins. Blood-borne humoral factors may alter the responsiveness of the blood vessels to metabolites and other control mechanisms.

Myogenic control is the result of the responsiveness of the vascular smooth muscle to changes in transmural pressure. Distention of an arterial vessel because of an increase in transmural pressure causes the vessel to constrict, whereas decreasing the transmural pressure causes the vessel to dilate. Myogenic control may be important for the regulation of both blood flow and fluid filtration.

The brain and the heart are examples of tissues in which neural control seems to be of minor importance under normal physiological conditions. Blood flow to both these organs is regulated by local control mechanisms. Changes in metabolism cause marked alterations in blood flow to these tissues. Cerebral blood vessels are extremely sensitive to variations in blood or cerebral spinal fluid carbon dioxide concentration, whereas metabolic control of coronary blood flow seems to occur primarily through changes in adenosine concentration.

Neural control has a much greater influence on the regulation of blood flow to the skin and resting skeletal muscle. In exercising skeletal muscle, however, local metabolic control overrides the influence of the sympathetic nervous system. The regulation of cutaneous blood flow under many circumstances is not directed to maintaining the metabolic needs of this tissue, but is more closely involved in temperature regulation of the body as a whole.

The regulation of gastrointestinal circulation is closely associated with the secretory and absorptive functions of the gut. Blood flow is greatly influenced by vasoactive hormones which are produced in this region. Although neural control has a larger influence over gastrointestinal blood flow than it does over cerebral or myocardial flow, there is an attenuation of the effect on the arterial vessels when sympathetic stimulation is continued for more than a few minutes. This phenomenon is called "autoregulatory escape." Venous vessels in the gastrointestinal circulation do not display autoregulatory escape.

Autoregulation, defined as the maintenance of a constant blood flow in the face of changes in perfusion pressure, is a phenomenon which seems to occur to some degree in most organs of the body. Autoregulation seems to occur as a result of both metabolic and myogenic control mechanisms. The relative contributions of these two mechanisms to autoregulation may differ among organs. Although the sympathetic nervous system may modulate the range of pressures over which an organ autoregulates, the presence of sympathetic nerves is not necessary for autoregulation to occur.

REFERENCES

Altura, B. M. (1978). Humoral, hormonal, and myogenic mechanisms in microcirculatory regulation including some comparative pharmacologic aspects of microvessels. In "Microcirculation" (G. Kaley and B. M. Altura, eds.), Vol. 2, pp. 431–502. Univ. Park Press, Baltimore, Maryland.

Altura, B. M., and Halevy, S. (1977). Cardiovascular actions of histamine. In "Histamine and Antihistaminics" (M. Rocha e Silva, ed.), Handbook of Experimental Pharmacology, Vol. 18, Part 2, pp. 1–39. Springer-Verlag, Berlin and New York.

Baez, S. (1968). Bayliss response in the microcirculation. *Fed. Proc., Fed. Am. Soc. Exp. Biol.* **27**, 1410–1415.

Baez, S., Feldman, S., and Gootman, P. (1977). Central neural influence on precapillary microvessels and sphincter. *Am. J. Physiol.* **2**, H141–H147.

Barcroft, H., and Dornhorst, A. C. (1954). Blood-flow response to temperature and other factors. In "Peripheral Circulation in Man" (G. E. W. Wolstenholme and J. S. Freeman, eds.), Ciba Foundation Symposium, pp. 122–131. Little, Brown, Boston, Massachusetts.

Barlow, T. E., Haigh, A. L., and Walder, D. N. (1958). A search for arteriovenous anastomoses in skeletal muscle. *J. Physiol. (London)* **143**, 80.

Barlow, T. E., Haigh, A. L., and Walder, D. N. (1959). Dual circulation in skeletal muscle. *J. Physiol. (London)* **149**, 18–19.

Barlow, T. E., Haigh, L., and Walder, D. N. (1961). Evidence for two vascular pathways in skeletal muscle. *Clin. Sci.* **20**, 367–385.

Bayliss, W. H. (1902). On the local reactions of the arterial wall to changes in internal pressure. *J. Physiol. (London)* **28**, 220–231.

Berne, R. M. (1963). Cardiac nucleotides in hypoxia: Possible role in regulation of coronary blood flow. *Am. J. Physiol.* **204**, 317–322.

Berne, R. M., and Levy, M. N. (1977). "Cardiovascular Physiology." Mosby, St. Louis, Missouri.

Bohlen, H. G., and Gore, R. W. (1979). Microvascular pressure in rat intestinal muscle during direct nerve stimulation. *Microvasc. Res.* **17**, 27–37.

Carrier, O., Jr., Walker, J. R., and Guyton, A. C. (1964). Role of oxygen in autoregulation of blood flow in isolated vessels. *Am. J. Physiol.* **206**, 951–954.

Cobbold, A., Folkow, B., Kiellmer, I., and Mellander, S. (1963). Nervous and local control of precapillary sphincters in skeletal muscle as measured by changes in filtration coefficient. *Acta Physiol. Scand.* **57**, 180–192.

Collins, G. M., and Ludbrook, J. (1967). Behavior of vascular beds in the human upper limb at low perfusion pressure. *Circ. Res.* **21**, 319–325.

Detar, R., and Bohr, D. F. (1968). Oxygen and vascular smooth muscle contraction. *Am. J. Physiol* **214**, 241-244.
Downey, J. M., Downey, H. F., and Kirk, E. S. (1974). Effects of myocardial strains on coronary blood flow. *Circ. Res.* **34**, 286-292.
Dresel, P., Folkow, B., and Wallenkin, I. (1966). Rubidium[86] clearance during neurogenic redistribution of intestinal blood flow. *Acta Physiol. Scand.* **67**, 173-184.
Duling, B. R., and Pittman, R. N. (1975). Oxygen tension: Dependent or independent variable in local control of blood flow? *Fed. Proc., Fed. Am. Soc. Exp. Biol.* **34**, 2012-2019.
Edvinsson, L. (1975). Neurogenic mechanisms in the cerebrovascular bed. *Acta Physiol. Scand., Suppl.* **427**, 5-35.
Eriksson, E., and Lisander, B. (1972). Changes in precapillary resistance in skeletal muscle vessels studied by intravital microscopy. *Acta Physiol. Scand.* **84**, 295-305.
Folkow, B. (1967). Regional adjustments of intestinal blood flow. *Gastroenterology* **52**, 423-432.
Fox, R. H., and Hilton, S. M. (1958). Bradykinin formation in human skin as a factor in heat vasodilation. *J. Physiol. (London)* **142**, 219-232.
Funaki, S. (1961). Spontaneous spike-discharges of vascular smooth muscle. *Nature (London)* **146**, 1102-1103.
Furness, J. B. (1973). Arrangement of blood vessels and their relation with adrenergic nerves in the rat mesentery. *J. Anat.* **115**, 347-364.
Furness, J. B., and Marshall, J. M. (1974). Correlation of the directly observed responses of mesenteric vessels of the rat to nerve stimulation and noradrenaline with the distribution of adrenergic nerves. *J. Physiol. (London)* **239**, 75-88.
Gore, R. W., and Bohlen, H. G. (1975). Pressure regulation in the microcirculation. *Fed. Proc., Fed. Am. Soc. Exp. Biol.* **34**, 2931-2937.
Heistad, D. D., and Abboud, F. M. (1974). Factors that influence blood flow in skeletal muscle and skin. *Anesthesiology* **41**, 139-156.
Heistad, D. D., and Marcus, M. L. (1978). Evidence that neural mechanisms do not have important effects on cerebral blood flow. *Circ. Res.* **42**, 295-302.
Johnson, P. C. (1960). Autoregulation of intestinal blood flow. *Am. J. Physiol.* **199**, 311-318.
Johnson, P. C. (1971). Gastrointestinal circulation. *In* "Physiology" (E. Selkurt, ed.), pp. 569-578. Little, Brown, Boston, Massachusetts.
Johnson, P. C. (1974). The microcirculation and local and humoral control of the circulation. *In* "International Review of Physiology, Vol. 1, Cardiovascular Physiology" (A. C. Guyton and C. E. Jones, eds.), pp. 163-195. University Park Press, Baltimore, Maryland.
Johnson, P. C., and Wayland, H. (1967). Regulation of blood flow in single capillaries. *Am. J. Physiol.* **212**, 1405-1415.
Kennedy, C. H., Grave, G. D., and Sokoloff, L. (1971-1972). Alterations of local cerebral blood flow due to exposure of newborn puppies to 80-90% oxygen. *Eur. Neurol.* **6**, 137.
Kontos, H. A. (1975). Mechanisms of regulation of the cerebral microcirculation current concepts of cerebrovascular disease. *Stroke* **10**, 7-12.
Kontos, H. A., Raper, A. J., Patterson, J. L., Jr. (1971). Mechanisms of action of CO_2 on pial precapillary vessels. *Eur. Neurol.* **6**, 114-118.
Korner, P. I., Chalmers, P., and White, S. W. (1967). Some mechanisms of reflex control of the circulation by the sympathoadrenal system. *Circ. Res.* **20/21**, Suppl. III, 157-172.
Krogh, A. (1929). "The Anatomy and Physiology of Capillaries." Yale Univ. Press, New Haven, Connecticut.
Lassen, N. A. (1978). Brain. *In* "Peripheral Circulation" (P. C. Johnson, ed.), pp. 337-358. Wiley, New York.
Lindbom, L., Tuma, R. F., and Arfors, K-E. (1980). Influence of oxygen on perfused capillary red cell velocity in rabbit skeletal muscle. *Microvasc. Res.* **19**, 197-208.

References

Lundgren, O. (1967). Studies in blood flow distribution and countercurrent exchange in the small intestine. *Acta Physiol. Scand., Suppl.* **303**, 3-42.

Lundgren, O. (1978). The alimentary canal. *In* "The Peripheral Circulation" (P. C. Johnson, ed.), pp. 255-284. Wiley, New York.

Lundgren, O., and Sranik, J. (1973). Mucosal hemodynamics in the small intestine. *Acta Physiol. Scand.* **88**, 551-563.

Martini, J., and Honig, C. L. (1969). Direct measurement of intercapillary distance in beating rat heart *in situ* under various conditions of O_2 supply. *Microvasc. Res.* **1**, 244-256.

Mohamed, M. S., and Bean, J. W. (1951). Local and general alterations of blood CO_2 and influence of intestinal motility in regulation of intestinal blood flow. *Am. J. Physiol.* **167**, 413-425.

Pawlik, W., Shepherd, A. P., and Jacobson, E. D. (1975). Effects of vasoactive agents on intestinal oxygen consumption and blood flow in dogs. *J. Clin. Invest.* **56**, 484-490.

Pittman, R. N., and Duling, B. R. (1973). Oxygen sensitivity of vascular smooth muscle: *In vitro* studies. *Microvasc. Res.* **6**, 202-211.

Purves, M. J. (1978). Do vasomotor nerves significantly regulate cerebral blood flow? *Circ. Res.* **43**, 485-491.

Rhodin, J. A. (1967). The ultrastructure of mammalian arterioles and precapillary sphincters. *J. Ultrastruc. Res.* **18**, 181-223.

Rubio, R., and Berne, R. M. (1969). Release of adenosine by the normal myocardium in dogs and its relationship to the regulation of coronary resistance. *Circ. Res.* **25**, 407-415.

Rubio, R., and Berne, R. M. (1978). Myocardium. *In* "Peripheral Circulation" (P. C. Johnson, ed.), pp. 231-254. Wiley, New York.

Schaper, W. (1971). "The Collateral Circulation of the Heart." North-Holland Publ., Amsterdam.

Shepherd, J. T. (1963). "Physiology of the Circulation in Human Limbs in Health and Disease." Saunders, Philadephia, Pennsylvania.

Smith, D. J., and Vane, J. R. (1966). Effects of oxygen tension on vascular and other smooth muscle. *J. Physiol. (London)* **186**, 284-294.

Sparks, H. V. (1978). Skin and muscle. *In* "Peripheral Circulation" (P. C. Johnson, ed.), pp. 193-231. Wiley, New York.

Tuma, R. F., Lindhom, L., and Arfors, K-E. (1977). Dependence of reactive hyperemia in skeletal muscle on oxygen tension. *Am. J. Physiol.* **233**, H289-H294.

Tuma, R. F., Childs, C. M., Intaglietta, M., and Arfors, K.-E. (1975). Microvascular flow pattern in the tenuissimus muscle. *Bibl. Anat.* **13**, 151-152.

Vanhoutte, P. M. (1978). Heterogeneity in vascular smooth muscle. *In* "Microcirculation" (G. Kaley and B. M. Altura, eds.), Vol. 2, pp. 181-309. Univ. Park Press, Baltimore, Maryland.

Walder, D. N. (1953). The local clearance of radioactive sodium from muscle in normal subjects and those with peripheral vascular disease. *Clin. Sci.* **12**, 153-167.

Walder, D. N. (1955). The relationship between blood flow, capillary surface area, and sodium clearance in muscle. *Clin. Sci.* **14**, 303-314.

Walder, D. N. (1968). Vascular pathways in skeletal muscle. *In* "Circulation in Skeletal Muscle" (O. Hudlicka, ed.), p. 101. Pergamon, Oxford.

Wiedeman, M. P. (1966). Contractile activity of arterioles in the bat wing during intraluminal pressure changes. *Circ. Res.* **19**, 559-563.

Wiedeman, M. P. (1968). Blood flow through terminal arterial vessels after denervation. *Circ. Res.* **22**, 83-89.

Wiedeman, M. P., Tuma, R. F., and Mayrovitz, H. N. (1976). Defining the precapillary sphincter. *Microvasc. Res.* **12**, 71-75.

Zweifach, B. W., and Metz, D. B. (1955). Regional differences in response of terminal vascular bed to vasoactive agents. *Am. J. Physiol.* **182**, 155-165.

6

Exchange in the Microcirculation

I. INTRODUCTION

The exchange of materials between the blood and tissues occurs primarily at the level of the capillaries. The capillaries are well designed for the exchange process with their large surface area and the structure of the capillary wall, which is composed of a single layer of endothelial cells and a basement membrane. There are differences in the structure of the capillary wall in different parts of the body. These differences in structure reflect modifications in the exchange process necessary for the function of specific organs.

Capillaries can be grouped into three general categories based on their ultrastructure; (1) fenestrated, (2) nonfenestrated, and (3) discontinuous. Fenestrated capillaries have circular openings in the endothelial cells with a diameter of 400–600 Å. These openings may be covered by a diaphragm or may actually be an open channel across the endothelial cell. Fenestrated capillaries are located in visceral organs, endocrine glands, the choroid plexus of the brain, and the cilliary body of the eye. Nonfenestrated capillaries are the most common type and are found in skin, connective tissue, skeletal and cardiac muscle, alveolar capillaries of the lung, and the brain. As implied by the name, there are no fenestrae or openings connecting the lumen and interstitium across the endothelial cell of the nonfenestrated capillaries.

Discontinuous capillaries are found in the liver, spleen, and bone marrow.

I. Introduction 141

These capillaries have wide lumens and large openings between the endothelial cells or in the endothelial cell itself. The morphology of these capillaries is greatly modified to meet the functional needs of the type of tissue in which they occur. Endothelial cells in the spleen and in the bone marrow are also highly phagocytic (Fig. 6.1).

Although the capillaries are considered to be the primary area of exchange, it should be pointed out that postcapillary venules are also involved in the exchange process. Like the capillaries, the postcapillary venule is composed of a single layer of endothelial cells and a basement membrane. Calculations from recently published data (Schmid-Schoenbein, 1976) show that the surface area of the postcapillary venules is even larger than that of the capillaries. The extreme thinness of the wall of the postcapillary venule and the large surface area facilitate exchange across this segment of the vasculature.

There are numerous pathways through which materials can move across the capillary wall. Matter and some solutes can move directly through endothelial cells. Cytoplasmic vesicles, which move between the lumen and the outer border of the endothelial cell, exchange fluid and solutes. There are intercellular junctions that permit diffusion and ultrafiltration and some that may be large enough to permit limited passage of plasma proteins. Intracellular fenestrations allow the passage of large molecules. Tight junctions, such as found in capillaries in the brain, are very restrictive to exchange. Large openings, such as sinusoids of the liver and the spleen, are easily traversed by all materials, including large proteins of the plasma (Fig. 6.2).

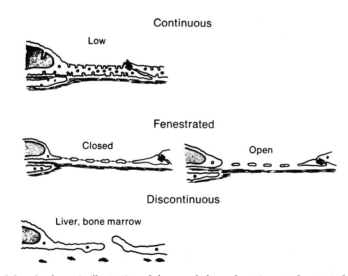

Fig. 6.1. A schematic illustration of the morphology of continuous, fenestrated, and discontinuous capillaries. (From Majno, 1965.)

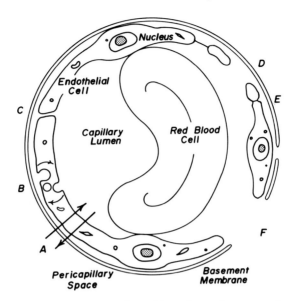

Fig. 6.2. A schematic illustration of possible pathways for transport across the capillary endothelium. A, transport directly through the endothelial cell; B, vesicular transport; C, transport through intercellular junctions; D, transport through intracellular fenestrations; E, transport through the tight junctions of the brain. (From Shepherd and Vanhoutte, 1979.)

The modes of transport through the morphological structures shown in the composite diagram are divided into three general categories; (1) diffusion, (2) filtration and osmosis, and (3) vesicular transport. Active transport by endothelial cells may occur in specialized areas but is not yet considered to be a primary mechanism.

II. DIFFUSION

Diffusion is the net transfer of molecules from areas of high concentration to areas of low concentration due to the random motions of individual molecules. The energy source for this random motion is the inherent kinetic energy of molecules. The rate of diffusion is proportional to the concentration gradient of the molecules, as described in the following equation.

$$J_s = -PS\,(C_i - C_o)$$

This states that the rate of movement, or flux, of a solute (J_s) is equal to the product of capillary permeability (P), capillary surface area (S), and the difference in concentration between the substance inside the capillary and the substance outside the capillary ($C_i - C_o$).

II. Diffusion

Applying these concepts to the vascular system, molecules which can freely pass across the capillary wall will diffuse into or out of the capillaries as long as there is a difference in concentration between the blood and the interstitial fluid.

The factors governing the rate of diffusion of oxygen from the capillaries to the interstitial space were first analyzed by Krogh (1919). From this analysis he showed that tissue oxygen tension is determined by tissue oxygen consumption, oxygen tension of the blood, and the distance between the capillaries that supply the tissue. The relationships are expressed in the following equation.

$$P_{T_{O_2}} = P_{C_{O_2}} - \frac{\dot{V}_{O_2}}{4\alpha D} \left[R^2 \ln\left(\frac{R}{r}\right)^2 - (R^2 - r^2) \right]$$

where

$P_{T_{O_2}}$ equals the minimum tissue P_{O_2}, $P_{C_{O_2}}$ equals the mean capillary P_{O_2}, $\dot{V}_{O_2}$ equals the oxygen consumption of the tissue, α equals the solubility of oxygen in the tissue, R equals the intercapillary distance, r equals the capillary radius, and D equals the diffusion coefficient for oxygen.

Although this equation does not completely express all of the factors governing oxygen transport in the tissues, it does show that tissue P_{O_2} can be increased in two ways; (1) by increasing the delivery of oxygen through increased blood flow and (2) by increasing the number of capillaries being perfused with blood. Except under conditions where total blood flow to an organ is very low, changing the intercapillary distance has a greater influence over tissue P_{O_2} than does changing the total blood flow. Therefore, normal or near normal levels of oxygen tension may be maintained even when there is a reduction in total blood flow to an organ if the number of perfused capillaries is increased. The converse is also true.

Until recently it was assumed that all significant exchange of oxygen occurred only in the capillary bed. However, Duling and Berne (1970) demonstrated by direct measurement in the hamster cheek pouch vessels that significant amounts of oxygen are lost from the blood of arterial vessels even before the capillaries are reached (Fig. 6.3).

The rate of diffusion of any molecule will of course, also depend on the permeability of the vessel wall to that molecule. In the early 1950s, Pappenheimer, Renkin, and Borrero attempted to characterize the permeability of capillaries to a number of different substances (Renkin, 1978). In these classic studies they found that capillary walls were extremely permeable to lipid-soluble materials. They also found that with increasing molecular weight, the permeability of the capillaries to the lipid-insoluble substances decreased more rapidly than one would predict from changes in their free diffusion coefficients. Finally, they were unable to measure permeability to lipid-insoluble substances when the molecular weight approached that of plasma proteins. As a result of these studies the investigators postulated the existence of small "pores" or water-filled channels between the blood vessel lumen and the

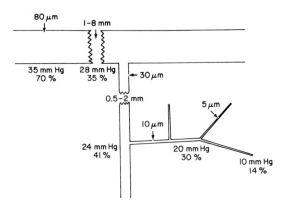

Fig. 6.3. A schematic illustration of changes in periarteriolar P_{O_2} in various vascular segments of the hamster cheek pouch. Perivascular P_{O_2} was found to agree closely with measurements of intravascular P_{O_2}. (From Duling and Berne, 1970, by permission of the American Heart Association, Inc.)

interstitum. They estimated that these pores were cylindrical channels with an equivalent radius of 30 Å or slits with a width of 40 Å. Later studies by Landis and Pappenheimer and others have placed the equivalent radius of the small pore between 40 and 50 Å (or a slit between 55 and 80 Å), with a density of $10-15/\mu m^2$ of endothelial luminal surface (Landis and Pappenheimer, 1963; Luft, 1965; Palade et al., 1979). Channels or pores of these dimensions restrict the size of hydrophilic molecules which could either be filtered or diffuse out of the capillaries. Since proteins and other macromolecules can move across the capillary wall, a "large pore" system was also postulated to exist. The diameter of the large pores is estimated to be between 500 and 700 Å, and their density is felt to be far lower than the density of the small pores (approximately 1 per 20 μm^2 of luminal surface in skeletal muscle capillaries.)

With the development of suitable techniques of electron microscopy it seemed reasonable that the morphological structures corresponding to both the large and small pores could be readily identified. However, at the present time the identification of these theoretical pores remains an area of intensive investigation and debate. From electron-microscopic studies of the transport of tracer material it has been postulated that the junction of endothelial cells is the site of the small pore (Karnovsky, 1967; Wissig, 1979). Other investigators, using the same techniques, conclude that the small pores are not located at the endothelial junction, but rather that vesicles represent both the small pore and large pore system. Wide endothelial clefts have also been reported to be the equivalent of the large pore.

To summarize, lipophilic substances such as oxygen and carbon dioxide and steroids are able to diffuse through endothelial cell membrane. These substances

III. Filtration and Osmosis

therefore have the entire endothelial surface area available for unrestricted exchange. Water, due to its small molecular size, is also able to diffuse through the endothelial cell and therefore has the entire endothelial surface area available for exchange. Unlike the unrestricted diffusion of lipophilic substances across the endothelium, hydrophilic substances such as glucose, amino acids, and proteins cannot diffuse through the endothelial cell membrane. These substances require aqueous channels such as pores for diffusion.

III. FILTRATION AND OSMOSIS

Contraction of the heart imparts a hydrostatic pressure to the blood in the aorta. Part of this hydrostatic pressure is transmitted to the blood in the capillaries. Since the capillaries are permeable to water, this hydrostatic pressure tends to force fluid out of the capillaries into the interstitium. Starling (1896) introduced one of the most important concepts in the area of exchange when he presented the idea that the filtration of fluid out of the capillaries would be opposed by the increasing colloid osmotic pressure of the blood, finally resulting in inward filtration of fluid at the venous end of the capillaries. Starling proposed that as fluid was forced out of the arterial side of the capillaries because of hydrostatic pressure, the concentration of proteins in the plasma would increase, since the capillary wall is relatively impermeable to these substances. The concentration difference between proteins in the plasma and the interstitial fluid creates an osmotic pressure gradient (the colloid osmotic or oncotic pressure) which tends to draw fluid back into the capillaries. He therefore felt that on the venous end of the capillary where hydrostatic pressure is relatively low and osmotic pressure is relatively high, inward filtration of fluid would occur. Landis (1927) verified Starling's hypothesis experimentally with measurements made on single capillaries of the frog mesentery. Starling's concept of fluid balance can be formulated by the following equation.

$$FM = K_r[(P_c - P_t) + (\pi_t - \pi_{pl})]$$

Where FM is the volume of fluid movement (either filtered or absorbed), K_F is the filtration coefficient (a measure of the permeability of the fluid across the capillary), P_c is the capillary hydrostatic pressure, P_t is the tissue hydrostatic pressure, π_t the interstitial colloid osmotic pressure, and π_{pl} the plasma colloid osmotic pressure. This formula simply states that the direction of fluid movement will be determined by the algebraic sum of the forces tending to move fluid out of the capillaries, i.e., the capillary hydrostatic pressure and the tissue colloid osmotic pressure, and also those factors tending to force fluid into the capillaries, i.e., the tissue hydrostatic pressure and plasma colloid osmotic pressure. This concept is illustrated in Fig. 6.4. The colloid osmotic pressure of the interstitial

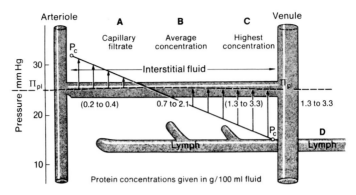

Fig. 6.4. A schematic illustration of the factors involved in fluid exchange across a single capillary. (From Landis and Pappenheimer, 1963.)

fluid occurs because of leakage of protein out of the capillaries into the interstitial fluid.

Although the Starling–Landis model of fluid exchange remains a useful guide in understanding the determinants of filtration across the capillary wall, quantitative measurements in the microcirculation have raised a number of questions about the Starling–Landis concept, especially when this concept is applied to individual capillaries. As stated previously according to the Starling–Landis concept, almost all of the fluid filtered out of the capillaries is returned to the vascular space by inward filtering on the venous side. This model does not take into consideration heterogeneity of capillary flow. Direct observation of the microcirculation shows that neither the morphology nor the perfusion of capillaries is homogeneous, even in individual tissue types. Therefore, it is possible that in some capillaries hydrostatic pressure may be very high, resulting in only outward filtration, whereas in other capillaries hydrostatic pressure may be very low, resulting only in reabsorption. Intaglietta and Zweifach (1974) have even questioned the basic conclusion drawn by Starling that most of the fluid filtered from the capillaries returns to the blood via reabsorption by the capillaries. In studies of single capillaries in the mesentery and omentum, these investigators found hydrostatic pressure in the venous end of the capillaries to be higher than plasma colloid osmotic pressure, and they only rarely observed reabsorption in this segment of the capillaries using the micro-occlusion technique. These investigators also state that the leakage of protein from the venous side of the capillaries and postcapillary venules may negate the colloid osmotic gradient predicted by Starling in this area. Wiederhielm (1968) has proposed that tissue colloid osmotic pressure may be sbustantially higher than would simply be predicted from the concentration of protein due to the effects of hyaluronic acid present in the tissue. A factor which also must be considered is that stated in the hypothesis by Guyton *et al.* (1971), i.e., the interstitial fluid pressure might actually be negative. All of these factors would contribute to outward filtration even at the venous end of the capillary. These findings lead Intaglietta and Johnson (1978) to conclude that in capillaries of many tissues, e.g., mesentery and renal glomerular capillaries, outward filtration occurs along the entire length of the capillary. In tissues such as skeletal and intestinal muscle, the capillary hydrostatic pressures are nearly equal to colloid osmotic pressure so that filtration and reabsorption according to the classical Starling–Landis model may occur. In areas such as the intestinal mucosa, reabsorption occurs along the entire length of the capillary. In tissues where outward filtration is the predominant direction of fluid movement, return of this fluid to the vascular space occurs primarily via the lymphatic system and not through reabsorption into the capillaries.

Thus, the Starling–Landis formulation is a very useful summary of the factors influencing fluid movement across the capillary wall. However, sufficient quantitative measurements of the components of the Starling–Landis formulation are not available at this time to predict precisely how much reabsorption of fluid occurs in the microvessels. Today, almost one-hundred years after Starling presented what he thought to be the principal mechanism of the return of fluid to the circulation, a number of investigators feel that the lymphatic system may be the primary pathway in many tissues, which was the accepted idea prior to Starling's work.

The amount of fluid actually filtered out of the capillaries is estimated to be less than 1–5% of the total amount of fluid passing through the capillaries in most organs (Crone, 1973; Intaglietta and Zweifach, 1974). Therefore, the movement of low molecular weight solutes due to filtration alone must be of little importance with regard to the total amounts of these molecules moving across the capillary (Crone, 1973). Diffusion is the mode of exchange primarily responsible for the movement of solute across the capillaries.

IV. VESICULAR TRANSPORT

One predominant feature of capillary endothelial cells is the presence of vesicles. Capillary endothelial cells contain a large number of vesicles which have been proposed to be involved in the transport of materials from the capillary lumen to the interstitium (Palade *et al.*, 1979). These vesicles range in size from 500 to 900Å and are surrounded by a membrane similar in appearance to the plasma membrane of the endothelial cell. The vesicles are primarily located in the narrow nonnucleated portion of the endothelial cell. The density of vesicles in endothelial cells varies between organs (the lowest density is in the brain) and also between segments of the microcirculation. The density is higher in capillaries than in venules, which have a higher density than arterioles. Within the capillary itself the density is highest on the venous end.

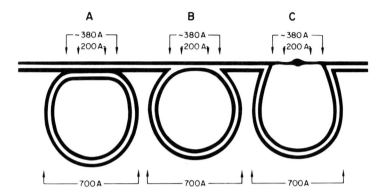

Fig. 6.5. A schematic illustration of the theoretical stages of attachment of vesicles with the plasma membrane of the endothelial cell. (From Palade et al., 1979.)

There is little doubt that vesicles are capable of transporting both macro- and micromolecules across the capillary endothelium. The vesicles are able to diffuse through the endothelial cytoplasm and attach to the endothelial plasma membrane on either the luminal or interstitial (antiluminal) side of the endothelial cell. It appears that following attachment of a vesicle with the endothelial plasma membrane, there is a progressive attenuation of the membranes at the point of attachment forming a single layer nonlipid stomatal diaphragm (Fig. 6.5). The fact that the stomatal diaphragm is not composed of lipids is important in that it appears to be permeable to lipid insoluble molecules up to an effective diameter of 50–110Å. Finally, as a continuation of this process, the diaphragm disappears and the vesicle is in open communication with the cell exterior. The diameter of the opening varies from less than 100 up to 400Å. During these processes, fluid and solute from the plasma become incorporated into the vesicles attached on the luminal side.

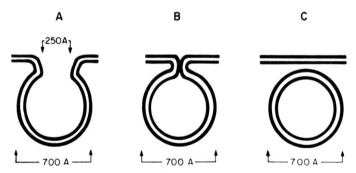

Fig. 6.6. A schematic illustration of the theoretical stages of detachment of vesicles from the plasma membrane. (From Palade et al., 1979.)

IV. Vesicular Transport 149

The process of vesicular detachment is illustrated in Fig. 6.6. Once detached, the vesicles then diffuse through the cytosol, and at least some of these vesicles will become attached to the plasma membrane on the opposite side of the endothelial cell where its contents are discharged. Under the conditions described in the cycle above, the vesicles could function as "large pores."

Electron micrographs have also demonstrated that in thin portions of the endothelial cell, a single vesicle may become attached to both the luminal and the interstitial side of the endothelial cell membrane. If the stomatal diaphragms are lost on both sides, this may constitute an open channel between the plasma lumen and the interstitium (Fig. 6.7).

Two or more vesicles may also fuse to form a chain, and this also may constitute an open channel between the plasma and the interstitium. The formation of open channels by either single vesicles or chains of vesicles may represent a second mode of vesicular transport. Open channels would represent a pathway open to both diffusive and connective transport (Fig. 6.8).

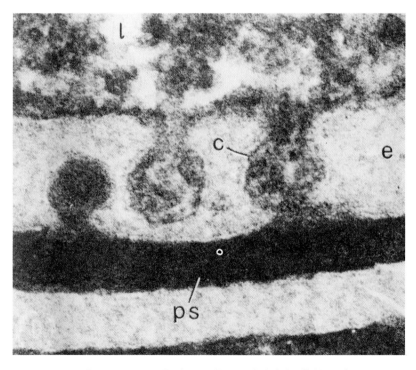

Fig. 6.7. An electronmicrograph of a capillary endothelial cell (e) in the rat cremaster muscle showing vesicle attached to both sides of the cell membrane and forming an open transendothelial channel (c) between the lumen of the capillary (l) and the pericapillary space (ps). (From Simionescu *et al.*, 1976.)

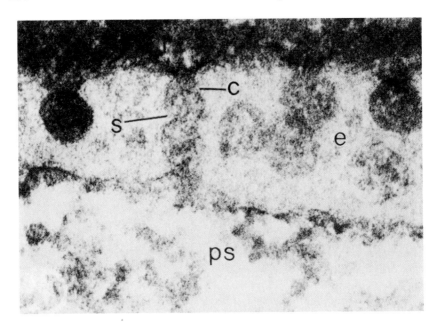

Fig. 6.8. An electronmicrograph of a patent transendothelial channel (c) formed by the fusion of vesicles in a capillary endothelial cell (e) in the rat diaphragm. This channel forms a direct connection between the capillary lumen (l) and pericapillary space (ps). Constrictions (s) at the fusion points of the vessels can be seen. (From Simionescu *et al.*, 1976.)

The vesicular chains forming transendothelial channels have constrictions at areas of fusion between vesicles, or between vesicles and the endothelial cell membrane, with diameters ranging from 400 to less than 100Å. Because of these constrictions it has been suggested that this is a pathway with size-restricted permeability to lipid-insoluble molecules and that this may be the morphological equivalent to the "small pores" in nonfenestrated capillaries (Simionescu *et al.*, 1975). These investigators have also proposed that fenestrations are produced by vesicles forming transendothelial channels, as illustrated below, and that fenestrations may also represent either "large" or "small pores" (Fig. 6.9).

According to this theory, the formation of transendothelial channels in nonfenestrated capillaries and the formation of fenestrae are simply variations of a common process. The stomated diaphragm of vesicles and the diaphragm of closed fenestratae appear similar in structure. In both cases, the diaphragm appears to be composed of protein, not lipid, which may allow limited passage of lipid-insoluble solute.

At the present time it is still not possible to make a definitive statement concerning the morphological equivalent of the "small pore." Using very similar techniques, Wissig (1979) and Palade *et al.* (1979) have come to opposite

V. Summary

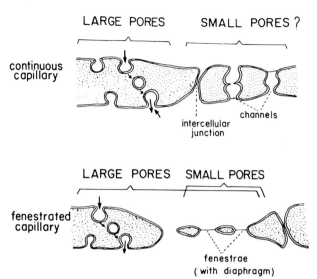

Fig. 6.9. A schematic diagram illustrating the possible morphological equivalents of the "large" and "small pores." (From Simionescu *et al.*, 1976.)

conclusions. Wissig feels that the vesicular channels are not of sufficient density to be the morphological equivalent of the small pore and that the endothelial junctions do, indeed, behave like the postulated small pores. Conversely, Palade *et al.* (1979) conclude that the permeability at the junctions of the endothelial cells is too restrictive for these junctions to be the morphological equivalents of small pores. These investigators feel that vesicular channels do occur with sufficient frequency to act as small pores. Until more quantitative data are available, and differences due to methods of preparation are resolved, the morphological equivalent of the small pore will remain uncertain.

V. SUMMARY

The exchange of materials between the blood and the tissues occurs primarily across the capillaries. Capillaries can be grouped into three general categories based on their ultrastructure; (1) nonfenestrated, (2) fenestrated, and (3) discontinuous. The differences in structure reflect modifications in the exchange process necessary for the function of specific organs. Among the possible pathways for movement of materials from blood to the interstitium are: movement directly through endothelial cells, through intercellular junctions, through intracellular fenestrations and transport by cytoplasmic vesicles. The modes of transport can be categorized as diffusion, filtration and osmosis, and vesicular transport.

Based on studies of the permeability of capillaries to various substances, two populations of pores have been postulated to exist. Small pores have been postulated to occur in substantially larger numbers than large pores and offer greater restriction to the transport of hydrophilic substances. The morphological structures corresponding to either the small or large pores remain to be identified.

REFERENCES

Crone, C. (1973). Capillary permeability: II. Physiological considerations. In "The Inflammatory Process" (B. W. Zweifach, L. Grant, and R. T. McCuskey, eds.), 2nd ed., Vol. 2, pp. 95–119. Academic Press, New York.

Duling, B. R., and Berne, R. M. (1970). Longitudinal gradients in periarteriolar oxygen tension. Circ. Res. 27, 669–678.

Guyton, A. C., Granger, H. J., and Taylor, A. E. (1971). Interstitial fluid pressure. Physiol. Rev. 51, 527–563.

Intaglietta, M., and Johnson, P. C. (1978). Principles of capillary exchange. In "Peripheral Circulation" (P. C. Johnson, ed.), pp. 141–166. Wiley, New York.

Intaglietta, M., and Zweifach, B. W. (1974). Microcirculatory basis of fluid exchange. Adv. Biol. Med. Phys. 15, 111–155.

Karnovsky, M. J. (1967). The ultrastructural basis of capillary permeability studied with peroxidase as a tracer. J. Cell Biol. 35, 213–236.

Krogh, A. (1919). The supply of oxygen to the tissues and the regulation of capillary circulation. J. Physiol. (London) 52, 457–474.

Landis, E. (1927). The capillary pressure in frog mesentery as determined by microinjection methods. Am. J. Physiol. 75, 548–570.

Landis, E. M., and Pappenheimer, J. R. (1963). Exchange of substances through the capillary walls. In "Handbook of Physiology, Circulation II" (W. F. Hamilton and P. Dow, eds.), Sect. 2, Vol. II, pp. 961–1034. Am. Physiol. Soc., Washington, D.C.

Luft, J. H. (1965). The ultrastructural basis of capillary permeability. In "The Inflammatory Process" (B. W. Zweifach, L. Grant, and R. T. McCluskey, eds.), pp. 121–159. Academic Press, New York.

Majno, G. (1965). Ultrastructure of the vascular membrane. In "Handbook of Physiology, Circulation III" (W. F. Hamilton and P. Dow, eds.), Section 2, Vol. 3, pp. 2293–2375. Am. Physiol. Soc., Washington, D.C.

Palade, G. E., Simionescu, M., and Simionescu, N. (1979). Structural aspects of the permeability of the microvascular endothelium. Acta Physiol. Scand., Suppl. 463, 11–32.

Renkin, E. M. (1978). The Microcirculatory Society Eugene M. Landis Award Lecture. Transport pathways through capillary endothelium. Microvasc. Res. 15, 123–135.

Schmid-Schoenbein, H. (1976). Microrheology of erythrocytes, blood viscosity, and the distribution of blood flow in microcirculation. In "Cardiovascular Physiology II" (A. C. Guyton, and A. W. Cowley, eds.) International Review of Physiology, Vol. 9, pp. 1–62. Univ. Park Press, Baltimore, Maryland.

Shepherd, J. T., and Vanhoutte, P. M. (1979). Components of the cardiovascular system: How structure is geared to function. In "The Human Cardiovascular System," p. 20. Raven, New York.

Simionescu, N., Simionescu, M., and Palade, G. E. (1975). Permeability of muscle capillaries to small hemepeptides: Evidence for the existence of patent transendothelial channels. J. Cell Biol. 64, 586–607.

References

Simionescu, N., Simionescu, M., and Palade, G. (1976). Structural–functional correlates in the transendothelial exchange of water-soluble macromolecules. *Thromb. Res.* **8,** Suppl. II, p. 257–269.

Starling, E. H. (1896). On the absorption of fluids from the connective tissue spaces. *J. Physiol (London)* **19,** 312–326.

Wiederhielm, C. A. (1968). Dynamics of transcapillary fluid exchange. *In* "Biological Interfaces. Flows and Exchanges" (F. P. Chinard, ed.), pp. 29–63. Little, Brown, Boston, Massachusetts.

Wissig, S. C. (1979). Identification of the small pore in muscle capillaries. *Acta Physiol. Scand.,* Suppl. **463,** 33–44.

HEMODYNAMICS

7

Quantitative Techniques for Measurement of Velocity and Pressure of Blood

I. MEASUREMENT OF BLOOD VELOCITY IN THE MICROCIRCULATION

Any method for the direct *in vivo* measurement of blood velocity in individual microvessels requires that the vessels of interest be directly observed through the microscope. Accuracy and limitations of all methods are dependent on a variety of interrelated factors, including (1) the method of illumination (transillumination or reflected light), (2) the type of illumination (continuous light or stroboscopic), (3) the size of the vessel, (4) the contrast of the observed target (red blood cell, plasma gap), (5) the magnification utilized, and (6) the actual velocity of the blood being measured. The general methods that have been used for the measurement of *in vivo* blood velocity may be conveniently divided into three general categories: direct or supplemented visual methods, which are for the most part limited to isolated velocity measurements and are principally of historic interest; photographic methods, which have utility under certain specialized conditions; and electro-optical methods, which are the techniques most frequently used at present when continuous velocity measurements are required.

7. Measurement of Velocity and Pressure of Blood

A. Visual Methods

One of the first descriptions of the measurement of red cell velocity in a living microcirculation has been attributed to Anton van Leeuwenhoek. The precise date of this account is somewhat in question, but it occurred somewhere in the late seventeenth century according to the translation offered by Hoole (1808). Van Leeuwenhoek assumed that he could pronounce a word of four syllables in about 0.83 seconds. Armed with this vocal timing mechanism he then observed the circulation of blood in an eel under the microscope and determined that in this time interval, the blood cells moved a distance of about 1.7 mm. Knowing the distance traveled by the red cell and the time it took, van Leeuwenhoek was able to calculate the velocity, which in this case was about 2 mm/sec. Though this method may appear crude, the numbers obtained were surprisingly accurate and, in fact, most methods presently used to measure blood velocity in small vessels are dependent on the determination of the transit time of the red cell or another observable particle over a known distance. It is probably evident that the reason van Leeuwenhoek's method worked is that he was able to identify either individual red cells or groups of cells. When this criterion is met, both visual and photographic methods can be used to obtain velocity information with certain limitations. From a practical viewpoint however, direct tracking of red blood cell movement is limited to about 1 mm/sec at a magnification of 500 and 2 mm/sec at a magnification of about 250 (Monro, 1966). These figures depend on the red cell concentration in the observed vessel as well as the contrast. The presence of plasma gaps, which sometimes have larger dimensions than the red cell, provide a larger "target"; consequently, under certain "granular" flow conditions, higher velocities than those quoted above may be visually determined. Several early workers used this visual tracking/stopwatch method and reported values for blood velocity in a variety of tissues (Hales, 1733; Weber, 1838). None of these values were larger than about 1.5 mm/sec, which is a consequence of the basic limitation of the visual method. Approaches to supplement the visual methods were initially constructed to help track the moving cell image by matching its movement with a controlled moving object. Such methods included the moving graticule technique described by Basler (1918) and later by Knisely (1934), the moving cathode ray spot of Bränemark (1959), and a mechanical moving spot device developed by Monro (1962). None of these methods permitted velocities above 2 mm/sec to be accurately determined, and each was somewhat cumbersome and gave only isolated velocity data. A novel technique capable of measuring velocities up to about 16 mm/sec was introduced by Monro (1964). Known as the streak-image technique, it relies on the fact that even though rapidly moving images are perceived by the eye as streaks, the retina can accurately measure the angle at which these streaks pass across it (Monro, 1966).

B. Photographic Methods

The basic concept intrinsic to photographic techniques is that if a photograph is taken of a red cell distribution, plasma gap, or cell group at a certain time, and at a subsequent time a second photograph is taken, then the movements of either a given cell or cell group can be identified and the velocity determined by calculation. Whether or not a photographic method is applicable to a given velocity measurement problem depends on a variety of factors.

One of the principal determinants is the minimum required framing rate. This minimum is dictated principally by two requirements: (1) the target being tracked must be in the microscopic field of view for at least two consecutive frames; and (2) there must be sufficient resolution to identify the target. The linear microscopic field is inversely related to the magnification used and to a small degree dependent on the specific details of the optical system used. The total field is not usable for photographic imaging because of resolution loss in peripheral zones. Further, the total image area, as recorded on the film plane, may be less than that visually observed unless specific effort to ensure a one to one correspondence is undertaken. Assuming that this is the case, Fig. 7.1 is a plot that shows the maximum linear field and the approximate useful field of view as a function of magnification used. As an example, it may be seen from this figure that to measure blood flow of 1 mm/sec using a magnification of 400× and the maximum field of view, the absolute minimum requirement is a frame rate of 2/sec. Higher magnifications and/or larger velocities require correspondingly faster rates. This framing rate is very difficult to achieve using a manual 35 mm camera and is close to the limit of motor drive 35 mm systems. Therefore, for most applications cinemicrophotography is required. Assuming that the photo-

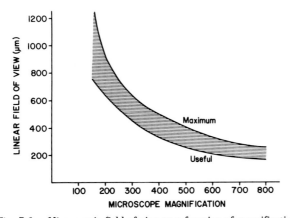

Fig. 7.1. Microscopic field of view as a function of magnification.

graphic system used can frame fast enough to capture the target image before it leaves the field of view, the second and more stringent requirement is that the "target" be identifiable—ideally, to freeze the image. To meet this criteria, blood cells moving at 1 mm/sec (1000 μm/sec) would require shutter speeds greater than 1/500 sec (35 mm auto drive), or framing rates of about 500/sec or greater to provide appropriate resolution under all but exceptionally optimum conditions and low red blood cell concentrations. Tracking of larger plasmatic gaps and/or blood cell groups impose less stringent conditions. Because of these considerations, the use of photographic methods for the measurement of RBC velocities is limited to very low velocities using conventional cine filming, or high velocities using very high framing rates.

Though numerous investigators have utilized high-speed cinephotography (up to 8000 frames/sec) to obtain velocity information, many technical problems must be overcome. Filming at the rate of 1000 frames/sec using 16 mm film is equivalent to about 60 ft of film/sec, which requires a large film magazine capability to obtain even a few seconds of data. Because of acceleration during start up, the initial film speed is not constant, and some type of timing mechanism is required to know the actual elapsed time between adjacent frames (Schlosser et al., 1965). Because of the low effective exposure times concomitant with high framing rates, a very high light intensity is required as well as appropriate heat absorbing filters to protect the preparation from intense heat; pulsed light, synchronous with the camera, may be used to reduce this problem (Monro, 1969). In spite of these difficulties, important *in vivo* data has been obtained using high-speed cinephotography (Bloch, 1962; Guest et al., 1963; Bond et al., 1965). In most cases, cine methods are restricted to applications in which the vessels are not significantly larger than capillary dimensions.

C. Electro-optical Techniques

Because of differences in optical density between RBC's and plasma, the optical image perceived by an observer viewing microvessel blood flow will vary in intensity. If, in place of the observer's eye, an opticoelectrical transducer (sensor) is used, then this temporal variation in opacity can be converted into an electrical signal. The amplitude of the signal will be proportional to the instantaneous light intensity falling on the photosensitive surface of the sensors. If the optics and sensor location and design of the system are chosen so that the sensor receives light only from a small segment of the vessel, then its electrical output may be thought of as a signal that is a temporally continuous representation of the blood flow patterns within a narrow spatial range. The signal represents a "signature" of the pattern at a particular spatial site as a function of time. When two such sensors are used and separated axially by a known distance such that one sensor monitors the upstream and the other the downstream signature, the resul-

I. Measurement of Blood Velocity in the Microcirculation

tant electrical signals may be used to determine blood velocity. Such determinations can be made visually, automatically, or may require supplemental signal processing, depending on the nature of the signals.

Visual determinations of blood velocity require that the **upstream** and **downstream** signatures be sufficiently similar, so that they are identifiable as having been caused by the same *in vivo* RBC pattern, but translated in the direction of flow and, hence, displaced in time.

Müller (1961), using the frog web, was the first to utilize photo-optic techniques to measure blood velocity in capillaries. Müller's method utilized a single CdSe photosensitive resistor positioned behind a 100 μm diameter pinhole in a screen upon which the microscopic image of the flowing RBC's was projected. Using accessory electronics, the time it took individual cells to pass the pinhole could be determined and translated into cell velocity. Wayland and Johnson (1967) refined this technique in two important ways. They incorporated a photomultiplier tube as their optical detector, following the lead of Wiederhielm *et al.* (1964), and more importantly, they used two slits and two sensors rather than one as employed by Müller. As a consequence, the transit time between cells could be determined. The basic ingredients of this so-called dual-slit method of velocity analysis is schematically represented in Fig. 7.2. Using this method, a right angle prism projects the microscopic image onto a screen that is penetrated by light sensing elements (light pipes leading to photomultiplier tubes, sensitive photo diodes, or transistors). The separation of the sensors depends on the magnification used, the velocity of the blood flow to be determined, and other factors that are

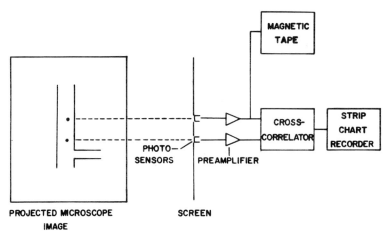

Fig. 7.2. Schematic representation of dual-slit velocity measuring technique. Microscope image is projected onto a viewing screen into which photosensors are inserted to detect changes in light intensity as blood flows past. The electrical signals from the photosensors are analyzed to determine the time delay between upstream and downstream patterns.

discussed below. Depending on the magnitude of the signal output of the photosensors and the desired cutoff frequency, preamplification may be required. The raw signal outputs may be recorded on magnetic tape for later analysis or printed out directly on a strip chart recorder. A cross correlator is employed to extract transit time information in more complex flow regimes in a manner which will now be described.

Figure 7.3 shows signal tracings obtained by Wayland and Johnson from mesenteric vessels. It should be noted that the signatures of the upstream and downstream detectors from the capillary and venule are readily identifiable visually, whereas the ability to make such a visual identification in the arteriole is much less obvious. When good signal variation is available, as it is within capillaries, threshold detection may be used to provide an automated determination of transit time between adjacent slits.

It is necessary to further process the signals to extract velocity information when the nature of the optical signatures is nondistinct, as is the case with larger

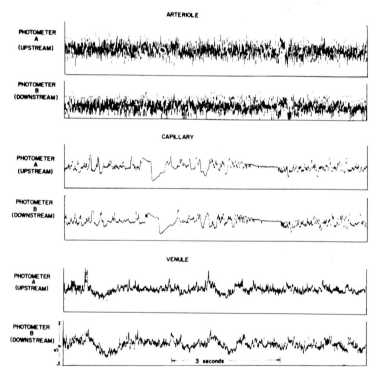

Fig. 7.3. Photosensor signals obtained from different microvessels. Note the similarity of the upstream and downstream patterns in the capillary and venule. (From Wayland and Johnson, 1967.)

I. Measurement of Blood Velocity in the Microcirculation

size microvessels and/or fast velocities. The principal method used is crosscorrelation. Analytically, crosscorrelation is a mathematical process which can be used to evaluate the degree of similarity between two signals. When applied to the determination of blood flow, the two signals are those obtained at each axially separated sensor site and are time-varying signals which may be denoted as $f_u(t)$ and $f_d(t)$, corresponding to the upstream and downstream signals, respectively. Under ideal conditions $f_d(t)$ would be identical in "shape" to $f_u(t)$, but simply displaced in time by an amount equal to the transit time δ between sampling sites.

Signals obtained from each finite size sensor (usually positioned in the center of the vessel image) have less than ideal similarity because of changes in cellular distribution during transit from one sensor to the other. The amount of dissimilarity depends on several factors, including blood velocity and vessel diameter, and is generally greater with increasing sensor separation. As long as this dissimilarity is not too great, the mathematical operation of crosscorrelation (defined by Eq. 1 below) can be used to determine the time lag, τ_{max}, at which the two signals are most similar. Since the sensor separation, δ (referred to the preparation), is known, velocity is calculable as δ/τ_{max}.

$$\phi_{ud}(\tau) = \lim_{T \to \infty} \frac{1}{T} \int_{-T}^{T} [f_u(t) f_d(t + \tau)] dt \tag{1}$$

The direct application of Eq. 1 to the measurement of blood velocity is not possible because the crosscorrelation function $\phi_{ud}(\tau)$ is defined over the entire history of the signal ($\infty \leq T \leq \infty$). In practice, a finite time interval T_0 is chosen over which to perform this mathematical operation. The start of the interval is arbitrarily taken as zero at the upstream sensor. Under these conditions the velocity that is calculated is the average velocity over the interval T_0. The appropriate value of T_0 is a compromise between being long enough to obtain sufficient data for reliable correlation and being small enough to follow dynamic changes in velocity.

Based on the pioneering work of Wayland and Johnson (1966), the correlation technique was refined and implemented as an on-line velocity measuring system by Intaglietta and co-workers (1970b). A block diagram of this system is shown in Fig. 7.4. The upstream and downstream signals, $V_u(t)$ and $V_d(t)$, were obtained from two photo diodes inserted into a screen so that the effective distance between them (referred to the preparation) was 4 μm. The time varying signals from each of these sensors is amplified in two matched AGC amplifiers. The outputs of each amplifier are routed to a special purpose correlation computer, which continuously determines the crosscorrelation, $C_{ud}(\tau)$, between the upstream and downstream signals. The peak value of this crosscorrelation is automatically detected and used to generate a pulse whose width is equal to the time lag corresponding to this maximum value (τ_{max}). A pulse width to voltage

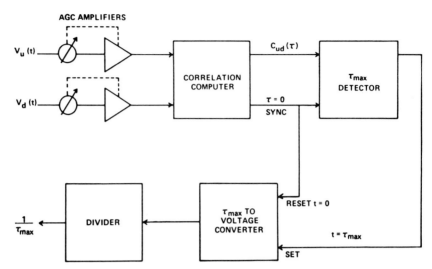

Fig. 7.4. Block diagram of an on-line RBC velocity measuring system. (From Intaglietta et al., 1970b.)

converter subsequently produces an output voltage that is proportional to τ_{max}. When this voltage is applied to an analog divider and subsequently multiplied by an appropriate scale factor, which takes into account the sensor separation, velocity is automatically determined. Continuous updating of changes in the value of τ_{max} and, hence, of velocity, is made possible because of the continuously computed correlation values. The frequency response of the system is a function of the time constant of the filters utilized, which provide temporary memory for the computed correlation function. Under ideal conditions, an output filter time constant of 100 msec yields a system frequency response of about 2.5 Hz.

As noted above, the degree of correlation between upstream and downstream signals decreases with sensor separation. This effect is illustrated in Fig. 7.5, which shows the correlation functions obtained in a 25 μm diameter arteriole for different sensor separations (Silva and Intaglietta, 1974). Position 1 shown in the figure denotes a sensor separation of 3.7 μm, and for each consecutive position, the sensor separation is increased by 1.6 μm. Three effects on the correlation function associated with increasing separation may be noted: (1) the position of peak shifts, since τ_{max} increases with the same velocity; (2) the peak that defines τ_{max} decreases; and (3) the function becomes progressively broader and less well defined. Analysis shows that for a given RBC velocity, the falloff in correlation is exponentially dependent on sensor separation, and that in 17–25 μm vessels, the degradation of the correlation is linearly related to the RBC velocity. From a practical point of view for the measurement *in vivo* blood velocities, it appears that sensor separations ranging between 6 and 18 μm may be employed for most

I. Measurement of Blood Velocity in the Microcirculation

purposes, although larger separations have application for the measurement of velocities less than about 2 mm/sec.

One of the important by-products of the measurement of blood velocities *in vivo* is its use to calculate volumetric flow rate in different vessel types under different physiological and/or pathological conditions. Because of the two phase structure of blood (RBC's and plasma), several difficulties are encountered when measurements of centerline RBC velocity (V_{CL}) are used to obtain volumetric flow. Consider the following illustrative example. Blood is caused to flow at a steady but unknown rate in a small tube. At the exit region of the tube the effluent is collected in a cup over a period of 100 sec. It is found that the total volume collected in the cup is 1 ml and that the volume fraction of RBC's in the cup is 0.4. From these determinations it is concluded that the total volumetric flow,

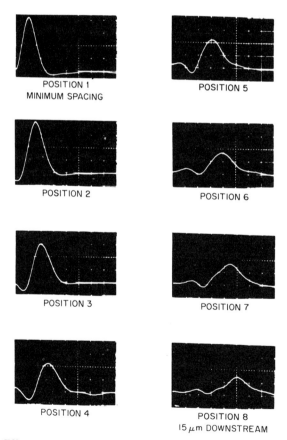

Fig. 7.5. Effect of photosensor separation on correlation between upstream and downstream signals. (From Silva and Intaglietta, 1974.)

(Q_T), is 1.0 ml/100 sec = 0.01 ml/sec; the volume flow of RBC's, Q_{rbc}, is 0.4 × 0.01 = 0.004 ml/sec; and the plasma flow, Q_{pl}, is 0.006 ml/sec. For each of these flows, a mean or bulk velocity may be defined which is simply the flow divided by an appropriate cross-sectional area, i.e., $V_T = Q_T/A$, where A is the total cross-sectional area of the tube. In order to predict this volumetric flow based on centerline velocity measurements, it is, in principle, necessary to know both the radial velocity and concentration profiles within the vessel. If, in the above example, the concentration profiles were uniform and the velocity profile was parabolic, with maximum value V_{max} in the tube center, then the mean velocity in the tube would be $V_{max}/2$, and the bulk flow, Q_T, (plasma plus RBC) would be equal to $V_{max}/2 \, (\pi D^2/4)$. For routine velocity measurements, however, neither the velocity nor concentration profiles are known accurately. Fortunately, experimental work carried out *in vitro* (Baker and Wayland, 1974; Lipowsky and Zweifach, 1978) have provided empirical constants that may be used to relate the measured centerline velocity to mean velocity and hence to the calculation of blood flow. Equation 2, below, expresses the volumetric flow as the product of the empirical constant α, the measured centerline V_{CL}, and the vessel cross-sectional area A.

$$Q = \alpha \, V_{CL} \, A \qquad (2)$$

For vessels having an internal diameter from 17 to 80 μm, the appropriate value of α is 0.625, a value confirmed by *in vitro* data obtained using dual-slit methods as well as image streaking techniques (Damon and Duling, 1979), and its constancy has been established *in vivo* (LaLone and Johnson, 1978). For vessels with diameters in the range of 80–140 μm, it has been suggested that V_{CL} is equal to V_{max} (Wayland, 1973), and therefore, an alpha value of 0.500 may be more appropriate. Finally, for vessels less than about 10 μm, an alpha value of 0.79 might be a more suitable approximation (Lipowsky and Zweifach, 1978), although the work of Starr and Frasher (1975a,b) indicate that the flow calculation in these smaller-sized vessels is more complicated.

The velocity measuring systems described above have the common feature of first converting the optical image of the flowing blood into an electrical signal. These converting arrangements are varied but currently utilize multiple photo transistors, which either intercept the light on a projection screen or are placed in specially designed housings on the microscope (Arfors et al., 1975). This latter arrangement eliminates the need for room darkness, which is a nuisance for some of the projected image methods. When video methods are applied to measurement of blood velocity, the television camera itself becomes the optical-to-electrical signal converter. The importance of closed circuit television (CCTV) as a tool in microcirculatory work is evidenced by the pioneering work of Bloch (1962, 1966), and by its use in automatically determining vessel diameters *in vivo*, as shown by Wiederhielm et al. (1964), Baez (1966, 1973), and Johnson et

I. Measurement of Blood Velocity in the Microcirculation

al. (1973), to mention only a few. When used to measure velocity as previously stated, the television camera serves as the primary optical–electrical converter. As differentiated from discrete photosensors, this method allows the entire image in the microscope to be viewed on a television monitor, stored on video tape or disk, and subsequently allows any part of the image to be analyzed and, if necessary, reanalyzed. Several methods of extracting the data to determine velocity are available, though one which derives from the dual-slit method has received the most attention. In this method, the video signal is routed through an electronic processing device, which has the ability to select any area of the monitored image and produce an output voltage that is proportional to the scene brightness within this selected area. The particular region of the image to be analyzed is indicated by a white cursor or "window," which is inserted into the video signal for identification purposes. The cursor size and position are controlled by the experimenter. To measure blood velocity, two cursors are used and bear the same relationship to the television image of a blood vessel as do photosensors to the projected image of the vessel in the dual-slit method. In principle, once the time-varying opacity signals that are derived from the upstream and downstream cursor locations are obtained, the method of crosscorrelation may be used to obtain velocity information. There are several significant system differences between these methods. The on-line dual-slit system acquires data continuously, whereas the television system uses discrete sampling and typically acquires data at a rate of 60 fields/sec (30 frames/sec) when standard U.S. systems are used. Persistence of the particular image tube used effectively reduces this theoretical maximum sampling rate. Since the spectral content of the photo-optic signals obtained on-line show significant components at frequencies well above this rate, considerable information is lost due to the discrete sampling process. Because of the presence of contaminating periodic signals associated with the 30 Hz framing rate, it is necessary to utilize low pass filters with sharp cut offs at about 20 Hz in order to obtain useful correlations. For standard systems, the highest velocity that can be routinely measured is about 1.5 mm/sec, although Fu and Lee in 1978 described a method whereby a video system was capable of velocity measurements of up to 12 mm/sec. In general, as a consequence of the limited velocity range that can be tracked by the video densimetric method, it has its principal application in the measurement of capillary events, including hematocrit and RBC velocity.

For any given application, careful thought to the type of television image tube is necessary. This includes the tube's sensitivity, noise level, and phosphor persistence. Many of these and other technical considerations have been previously reviewed (Bloch, 1966; Intaglietta *et al.*, 1975). The significant applications of television video densimetric methods include their use in experimental animals for the measurement of blood velocity and hematocrit in capillaries (Johnson and Wayland, 1967; Klitzman and Duling, 1979; Gussis *et al.*, 1979) and their extension to measurements in capillaries of the human nailfold (Bollinger *et al.*, 1974; Butti *et al.*, 1975; Fagrell, *et al.*, 1977) and of the human conjunctival vessels (Mayrovitz *et al.*, 1980, 1981). A new and promising non-invasive technique used to measure blood velocity in microvessels employs a Laser–Doppler principle. In this method, coherent laser light in the visible spectrum is focused on a vessel in which velocity is to be determined. A portion of the incident light is scattered from moving RBC's and is detected by an appropriate photosensitive device (photomultiplier). For a fixed wavelength of incident light and a given receiving angle, this scattered light will be shifted in frequency by an amount which is proportional to the velocity of the moving scattering objects. Measurement of the frequency shift forms the basis of the velocity determination. In practice, two components of scattered light are simultaneously detected; one is from the nonmoving structures (i.e., vessel wall), and the second is from the moving objects within the vessel. The scattered light from the nonmoving target will have a frequency exactly equal to that of the incident light, whereas that which is scattered from the moving red cells will have a frequency different from this by the amount of the Doppler shift. Both of these components are simultaneously received by the detector. The interaction of these two components produces a modulation in the intensity of the detected light, which consequently fluctuates at a frequency equal to the frequency shift produced by the moving object. Measuring frequency differences in this way is known as optical heterodyning detection. When light is scattered from the entire vessel cross section, a spectrum of frequency shifts are present in the modulated signal in an amount which depends on the velocity profile in the vessel. This fact requires that the detector signals be analyzed using some form of spectrum analysis to extract and display the frequency information. Riva *et al.* (1972) have argued on analytical grounds that if light incident on a vessel illuminates all flowing particles equally, and if the velocity profile is parabolic, then the power spectral density of the scattered light will be uniform from zero shift to a maximum, f_{max}, corresponding to the maximum velocity, V_{max} in the vessel. Application of the experimental technique to a dilute suspension of polystyrene spheres in 100 μm diameter capillary tubing produced a flat spectrum with a sharp cut off, in accordance with theory. Blood flowing in similar tubes resulted in a detected frequency spectrum which increased with frequency and rolled off slowly rather than having a sharp cut off. These characteristics may be explained on the basis of a flattened velocity profile in the tube and multiple scattering phenomena. The application of Laser–Doppler velocimetry (LDV) to the measurement of blood velocity in retinal vessels of anesthetized rabbits yielded reasonable velocity values and represents one of the first *in vivo* applications of this technique. Feke and Riva (1978) modified a fundus camera and applied the LDV technique to measure blood velocity changes in retinal veins (D = 150 μm) and arteries (D = 80 μm) in humans. Since, in their system, accurate determination of the scattering angle was not possible, absolute values of velocity could not be obtained. However, measurements in veins showed that the velocity was relatively constant, whereas in the artery, the velocity varied with the cardiac cycle. In each of these cases, utilization of the maximum frequency of the Doppler shift was utilized to estimate the maximum velocity in the vessel. The possibility of coupling incident and scattered light using fiber optics was investigated by Powers and Frayer (1978), who demonstrated frequency shifts sufficient to make RBC flow measurements *in vitro* in the skin of the human hand. Recent refinements in the Laser–Doppler method, including the use of interference fringes of parallel laser beams and automatic tracking of the Doppler-shift frequencies (Le-Cong and Zweifach, 1979), may portend a new era in obtaining microcirculatory human hemodynamic information.

II. MEASUREMENT OF BLOOD PRESSURE IN THE MICROCIRCULATION

One of the first reported attempts to measure changes in blood pressure in the microcirculation is attributed to Kreis (1875). Using glass plates as small as 2.5 mm^2 placed on a finger directly behind the nail, Kreis determined the amount of weight necessary to produce a distinct whitening of the underlying tissue. The weight to achieve this, divided by the surface area of the plate, was used as an index of the capillary pressure. Roy and Brown (1879) employed a much improved compression system and applied it to micropressure measurements in the web of the frog. In their method, a highly deformable transparent membrane prepared from calf peritoneum (and used at that time as a perfume bottle stopper) was secured to a cylinder and forced against the underlying vessels using controlled air pressure. Microvessel blood pressure was taken as that value of air pressure required to stop blood flow in the observed vessel. These investigators were among the first to report the presence of pulsatile pressure in the microcirculation: "As the extravascular pressure is raised, the blood flow through the capillaries becomes more and more pulsatile in character." They were also the first to describe, on somewhat quantitative grounds, the intrinsic temporal and spatial hemodynamic variability within the microcirculation: "... at one measurement we may find that, of two capillaries lying side by side, one will collapse with a pressure (18 mm Hg), while its neighbor continues to convey blood until a pressure of perhaps (26 mm Hg) is applied; ... on again raising the pressure the order in which they collapse may be reversed." Hill (1921) used this same apparatus to measure pressures in the wing vasculature of anesthetized bats and reported values ranging from 50 mm Hg in the main arterial branch to 15 mm Hg in arterioles. Though ingenious, these compression techniques have obvious limitations.

The era of direct microvascular pressure measurement opened in 1926 when Landis inserted finely drawn microcannula into mesenteric microvessels of the frog. Using dye-filled microcannula with tips between 4 and 8 μm in diameter, he observed the dye–plasma interface in the tip and adjusted the pressure in the cannula so that the interface did not move. Since a stationary interface is obtained only when the cannula pressure is equal to the intravascular pressure, Landis was able to obtain average microvascular pressures in this manner. By slowly increasing the pressure in the cannula by altering the external reservoir, diastolic and systolic pressures were estimated by noting the pressure at which the dye initially spurted out of the tip (diastolic pressure) and the point at which flow became continuous (systolic pressure). Nicoll (1969) much later used a similar backpressure technique to determine pressure in the bat wing microvasculature.

All methods currently used for direct measurement of blood pressure within individual microvessels require penetration of the vessel or its side branch by some type of microcannula. Although a variety of technical difficulties are associated with the preparation of these microcannula and their *in vivo* utilization (Bloch, 1966; Landis, 1966), the capabilities of this technique are dependent on more fundamental considerations. Although several passive transducer systems have been successfully used to obtain pressures in larger size microvessels (Rapaport *et al.*, 1959; LeVasseur *et al.*, 1969), a major problem is the limitation of the standard hydraulically coupled measuring system to adequately record temporal changes in intravascular pressure (Intaglietta, 1973; Wunderlich and Schnermann, 1969). This shortcoming arises because the measuring system behaves as a low-pass filter with a time constant that is inversely related to the fourth power of the tip diameter. Since the advent of solid state pressure transducers and their associated smaller compliance, this problem has been somewhat lessened. However in spite of a low value of system compliance (C), the high value of hydraulic resistance (R) at sufficiently small tip diameters produces a time constant (RC) which precludes faithful recording of all but the slowest changes in pressure.

To illuminate these points further, consider the following example. Blood pressure is to be measured in a 15 μm diameter arteriole. Glass microcannula are prepared using state of the art techniques (Fox and Wiederhielm, 1973), filled with appropriate fluid, placed into a micromanipulator, and directly coupled to a low compliance pressure transducer. The tip of the microcannula is positioned over the arteriole, and the vessel is penetrated. In order for the intravascular pressure to be detected, the transducer diaphragm must be displaced so that an appropriate electrical signal may be obtained. Since the lowest volume displacement transducer commercially available is presently rated at 6×10^{-9} cm³/mm Hg (corresponding to a compliance of 4.5×10^{-12} cm⁵/dyne), the volume of the closed hydraulic measuring system (microcannula plus transducer) must change by 6×10^{-9} cm³ to register a 1 mm Hg change in pressure. In addition to

II. Measurement of Blood Pressure in the Microcirculation

the transducer compliance, the nonzero compressibility of the fluid in the microcannula must be considered. Water at 35°C has a compressibility (K_W) of 4.5 × 10^{-11} cm²/dyne and the effective compliance of the microcannula due to water compressibility is $K_W V_W$, where V_W is the volume of fluid in the cannula. Minimum practical microcannula lengths are about 5 cm, and diameters are usually about 0.8 mm. The minimum cannula volume V_W is therefore of the order of 0.03 cm³, though practical coupling arrangements could easily require a total volume ten times as large as this. The minimum compliance then, due to water, is on the order of 1.4 × 10^{-12} cm⁵/dynes, and the total (minimum) compliance is about 6 × 10^{-12} cm⁵/dynes (8 × 10^{-9} cm³/mm Hg). This is the absolute minimum amount of fluid volume which must be transferred between cannula and blood vessel to register a 1 mm Hg change in blood pressure. It is this requirement for fluid transfer that gives rise to the difficulties. First and most obvious is that if the amount of fluid required to be transferred is a significant fraction of the flow in the vessel, then the presence of the cannula modifies the pressure that it is attempting to measure. In general, the smaller the vessel, the more serious this problem. However, even if the blood flow in the vessel is large enough so that the absolute amount of fluid transferred is not a problem, the rate at which this fluid can be transferred may be a serious limitation. If the hydraulic source resistance of the vessel supplying the fluid to the microcannula through the tip is assumed zero, then, as previously stated, the rapidity with which intravascular pressure changes are reflected by displacement of the pressure transducer depends on the hydraulic resistance of the microcannula and the compliance of the measuring system. The principal resistance is due to the tip, which for a typical length of 10 μm and a diameter of 1 μm is about 4 × 10^{12} dyne sec cm⁵. Using these numbers, the smallest conceivable time constant for this 1 μm tip passive measuring system is about 6 seconds—a value much too large to reliably follow pressure changes synchronous with heart rate or even changes due to normal vasomotion. However, since the frequency response is crucially dependent on the tip diameter, calculations show that for a heart rate of 120/min, accurate response can be obtained (20 msec time constant, cut-off frequency = 8 Hz) with a tip diameter of about 4.5 μm. A larger system fluid volume (say 0.3 versus 0.03 cm³) effectively increases the system compliance due to water by a factor of 10 and requires a tip diameter of about 8 μm to meet these same requirements in the present example.

Though the minimum tip diameters calculated above are slightly less than those reported for systems with larger transducer compliances (Intaglietta, 1973), it is still clear that for many applications they are too large, and alternatives to passive systems are required. One such system, introduced by Wiederhielm and co-workers (1964), extended the interface nulling technique of Landis in several important ways. First, rather than filling the microcannula with a dye so as to visually track fluid movement (or lack of that movement) within the tip, changes

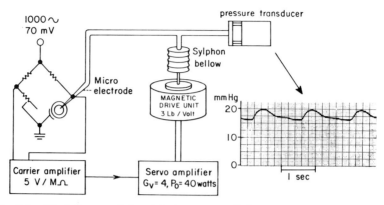

Fig. 7.6. Block diagram of the original servocontrolled micropressure measuring system. Waveform shown is that recorded from an arteriole in the frog mesentery. (From Wiederhielm et al., 1964.)

in electrical resistance of the tip are used. This is accomplished by filling the microcannula with a salt solution that has an electrical resistivity much less than that of the plasma in the microvessel in which the tip is inserted. Movement of fluid into or out of the tip will alter the electrolyte concentration profile within the tip and hence change its electrical resistance. This resistance change can be detected and converted into a proportional error signal by making the tip resis-

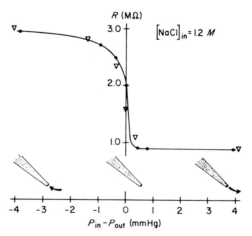

Fig. 7.7. Static resistance versus pressure characteristic of a 0.3 μm microcannula filled with 1.2 M NaCl. P_{in} and P_{out} are the pressures inside and outside the cannula, respectively. Note the high value of dR/dP in the narrow operating range thus ensuring high sensitivity and equivalency of inside and outside pressures when null is achieved. (From Fox and Wiederhielm, 1973.)

II. Measurement of Blood Pressure in the Microcirculation

tance one arm of a Wheatstone bridge. The second major feature of this system is its ability to develop an appropriate external counterpressure automatically rather than manually. This is accomplished by using the electrical signal change produced by tip fluid movement to drive an external pressure source connected to the microcannula. The effect of this external source is to minimize fluid movement in the tip while rapidly developing a counterpressure equal to the intravascular pressure. Figure 7.6 illustrates the original arrangement used by Wiederhielm. In practice, an AC Wheatstone bridge is used and is initially balanced so that a null voltage is provided to the carrier amplifier when the tip resistance is approximately midway in its operating range. Insertion of the tip into a blood vessel will tend to cause plasma entry into the tip due to the initial pressure gradient. With the first bit of plasma flow, the bridge becomes unbalanced because of the increased resistance of the tip; the resultant error signal is amplified and used to drive a bellows, which raises the pressure in the microcannula (microelectrode). This process continues until the counterpressure is approximately equal to the intravascular pressure (recorded by a standard pressure transducer), and the null condition of the bridge is restored. The operation of this servocontrolled pressure measuring system depends on the design parameters of the closed loop control system and the microcannula characteristics. To follow rapid changes in intravascular pressure, small changes in the concentration profile must be accompanied by large changes in tip resistance. That these conditions are indeed satisfied has been shown theoretically and experimentally by Firth and DeFelice (1971) and Fox and Wiederhielm (1973). Figure 7.7 illustrates the steep gradient

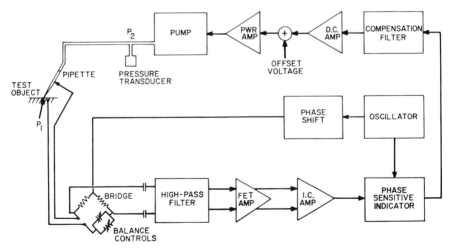

Fig. 7.8. Block diagram of an improved micropressure system. Current system incorporates a negative capacitance preamplifier to neutralize microelectrode capacitance, a band pass filter to allow for a 500 Hz bridge operation while eliminating 60 and 120 Hz noise, and a compensation filter in the feedback loop to stabilize the system up to a 35 Hz operation. (From Intaglietta et al., 1970a.)

in resistance accompanying small changes in pressure and the narrow operating range, thus ensuring the equality of inside and outside pressures.

The original system developed in 1964 was reported to have a frequency response flat to 20 Hz using tips from 0.5 to 5.0 μm. The use of smaller diameter tips (less than 0.1 μm) are accompanied by greater technical problems, including tip plugging, greater sensitivity to temperature and plasma composition, and their significant influence over contact potentials, (Fox and Wiederhielm, 1973); however, they are sharp enough for any purpose and have a frequency response more than adequate for the task. A variety of system improvements illustrated in Fig. 7.8 (Intaglietta *et al.*, 1970a; Intaglietta, 1973) and *in vivo* and *in vitro* tests have established this method of micropressure measurement as a most useful physiological tool.

REFERENCES

Arfors, K. E., Bergqvist, D., Intaglietta, M., and Westergren, B. (1975). Measurements of blood flow velocity in the microcirculation. *Uppsala J. Med. Sci.* **80**, 27–33.

Baez, S. (1966). Recording of microvascular dimensions with an image-splitter television microscope. *J. Appl. Physiol.* **21**, 299–301.

Baez, S. (1973). A method for in-line measurement of lumen wall of microscopic vessels *in vivo*. *Microvasc. Res.* **5**, 299–308.

Baker, M., and Wayland, H. (1974). On-line volume flow rate and velocity profile measurements for blood in microvessels. *Microvasc. Res.* **7**, 131–143.

Basler, A. (1918). Uber die blutbewegung in den kapillaren. *Pfluegers Arch. Gesamte Physiol. Menschen Tiere* **171**, 134–145.

Bloch, E. H. (1962). A quantitative study of the hemodynamics in the living microvascular system. *Am. J. Anat.* **110**, 125–153.

Bloch, E. H. (1966). Low-compliance pressure gauge. *Methods Med. Res.* **11**, 190–194.

Bollinger, A., Butti, P., Barras, J. P., Trachsler, H., and Siegenthaler, W. (1974). Red blood cell velocity in nailfold capillaries of man measured by a television microscopy technique. *Microvasc. Res.* **7**, 61–72.

Bond, T. P., Derrick, J. R., and Guest, M. M. (1965). High speed cinematograph studies of the microcirculation during hypothermia. *Bibl. Anat.* **7**, 191–194.

Bränemark, P. I. (1959). Vital microscopy of bone marrow in rabbit. *Scand J. Clin. Lab. Invest.*, *Suppl.* **38**.

Butti, P., Intaglietta, M., Reiman, H., Hollige, C., Bollinger, A., and Anliker, M. (1975). Capillary red blood cell velocity measurements in human nailfold by videodensitometric method. *Microvasc. Res.* **10**, 220–227.

Damon, D. N., and Duling, B. R. (1979). A comparison between mean blood velocities and center-line red cell velocities as measured with a mechanical image streaking velocitometer. *Microvasc. Res.* **17**, 330–332.

Fagrell, B., Fronek, A., and Intaglietta, A. (1977). Microscopic television system for studying flow velocity in human skin capillaries. *Am. J. Physiol.* **233**, H318–H321.

Feke, G. T., and Riva, C. E. (1978). Laser Doppler measurements of blood velocity in human retinal vessels. *J. Opt. Soc. Am.* **68**, 526–531.

Firth, D. R., and DeFelice, J. J. (1971). Electrical resistance and volume flow in glass microelectrodes. *Can. J. Physiol. Pharmacol.* **49**, 436–477.

References

Fox, J. R., and Wiederhielm, C. A. (1973). Characteristics of the servo-controlled micropipette pressure system. *Microvasc. Res.* **5**, 324–335.

Fu, S. F., and Lee, J. S. (1978). A video system for measuring the blood flow velocity in microvessels. *IEEE Trans. Biomed. Eng.* **25**, 295–291.

Guest, M. M., Bond, T. P., Cooper, R. G., and Derrick, J. R. (1963). Red blood cells: Change in shape in capillaries. *Science* **142**, 1319.

Gussis, G. L., Jamison, R. L., and Robertson, C. R. (1979). Determination of erythrocyte velocities in the mammalian inner renal medulla by a video velocity-tracking system. *Microvasc. Res.* **18**, 370–383.

Hales, S. (1733). "Statistical essays: Containing haemastatistics; or, an account of some hydraulic and hydrostatical experiments made on the blood and blood-vessels of animals." Vol. II. Innys, Manby & Woodward, London.

Hill, L. (1921). The pressure in the small arteries, veins and capillaries, of the bat's wing. *J. Physiol. (London)* **54**, 135.

Hoole, S. (1808). "Select Works of Anthony van Leeuwenhoek," Vol. 2, p. 344. Philanthropic Soc.

Intaglietta, M. (1973). Pressure measurements in the microcirculation with active and passive transducers. *Microvasc. Res.* **5**, 317–323.

Intaglietta, M., and Tompkins, W. R. (1971). Micropressure measurement with 1μ and smaller cannulae. *Microvasc. Res.* **3**, 211–214.

Intaglietta, M., Pawula, R. F., and Tompkins, W. R. (1970a). Pressure measurements in the mammalian microvasculature. *Microvasc. Res.* **2**, 212–220.

Intaglietta, M., Tompkins, W. R., and Richardson, D. R. (1970b). Velocity measurements in the microvasculature of the cat omentum by on-line method. *Microvasc. Res.* **2**, 462–473.

Intaglietta, M., Silverman, N. R., and Tompkins, R. W. (1975). Capillary flow velocity measurements in vivo and in situ by television methods. *Microvasc. Res.* **10**, 165–179.

Johnson, P. C., and Wayland, H. (1967). Regulation of blood flow in single capillaries. *Am. J. Physiol.* **212**, 1405–1415.

Johnson, P. C., Blascitke, J., Burton, K. S., and Dial, J. H. (1971). Influence of flow variations on capillary hematocrit in mesentery. *Am. J. Physiol.* **22**, 105–111.

Johnson, P. C., Hudnall, D. L., and Dial, J. H. (1973). Measurement of capillary hematocrit by photometric techniques. *Microvasc. Res.* **5**, 351–356.

Klitzman, B., and Duling, B. R. (1979). Microvascular hematocrit and red cell flow in resting and contracting striated muscle. *Am. J. Physiol.* **237**, H481–H490.

Knisely, M. H. (1934). Apparatus for illuminating living tissue and measuring rate and volume of blood flow. *Anat. Rec.* **58**, 73.

Kreis, N. V. (1875). Veber den Druk in den blutcapillaren der menschlichen haut. *Ber. Dtsch. Saechs. Ges. Wiss.*

LaLone, B. J., and Johnson, P. C. (1978). Estimation of arteriolar volume flow from dual slit red cell velocity: An *in vivo* study. *Microvasc. Res.* **16**, 327–339.

Landis, E. M. (1926). The capillary pressure in frog mesentery as determined by micro-injection methods. *Am. J. Physiol.* **75**, 548–570.

Landis, E. M. (1966). Microinjection pressure measuring technic. *Methods Med. Res.* **11**, 184–189.

Le-Cong, P., and Zweifach, B. W. (1979). *In vivo* and *in vitro* velocity measurements in microvasculature with a laser. *Microvasc. Res.* **17**, 131–144.

LeVasseur, J. E., Funk, F. C., and Patterson, J. L. (1969). Physiological pressure transducers for microhemocirculatory studies. *J. Appl. Physiol.* **27**, 422–425.

Lipowsky, H. H., and Zweifach, B. W. (1978). Application of the "two slit" photometric technique to the measurement of microvascular volumetric flow rates. *Microvasc. Res.* **15**, 93–101.

Mayrovitz, H. N., Duda, G., and Larnard, D. (1980). A video system to measure blood velocity in human conjunctival vessels. *Microvasc. Res.* **20**, 119.

Mayrovitz, H. N., Larnard, D., and Duda, G. (1981). Blood velocity measurement in human conjunctival vessels. *Cardiovasc. Dis.* (in preparation).

Monro, P. A. G. (1962). Measurement of blood cell velocity. *Bibl. Anat.* **1,** 110–115.

Monro, P. A. G. (1964). Visual particle velocity measurement: For fast particles and blood cells *in vivo* and *in vitro*. *Bibl. Anat.* **4,** 34–45.

Monro, P. A. G. (1966). Methods for measuring the velocity of moving particles under the microscope. *Adv. Opt. Electron Microsc.* **1,** 1–40.

Monro, P. A. G. (1969). Progressive deformation of blood cells with increasing velocity of flowing blood. *Bibl. Anat.* **10,** 99–103.

Muheim, M. H. (1977). Fabrication of glass micropipettes with well-defined tip configuration by hydro fluoric acid etching. *Bibl. Anat.* **16,** 348–350.

Müller, H. (1961). Uber eine methode zur durchstromungsregistrierung in einzelln kapillarein. *Z. Biol.* **113,** 39–51.

Nicoll, P. A. (1969). Intrinsic regulation in the microcirculation based on direct pressure measurements: *In* "Microcirculation" (W. L. Winters and A. N. Brest, eds.), pp. 89–101. Thomas, Springfield, Illinois.

Powers, E. W., and Frayer, W. W. (1978). Laser Doppler measurement of blood flow in the microcirculation. *Plast. Reconstr. Surg.* **61,** 250–255.

Rappaport, M. D., Bloch, E. H., and Irwin, J. W. (1959). A manometer for measuring dynamic pressures in the microvascular system. *J. Appl. Physiol.* **14,** 651–655.

Riva, C., Ross, B., and Bender, G. B. (1972). Laser Doppler measurement of blood flow in capillary tubes and retinal arteries. *Invest. Ophthalmol.* **11,** 936–944.

Roy, J., and Brown, J. G. (1879). The blood pressure and its variations in the arterioles, capillaries, and smaller veins. *J. Physiol. (London)* **2,** 323–359.

Schlosser, D., Heyse, E., and Bartels, H. (1965). Microcinematographic measurement of erythrocyte flow rate in lung capillaries. *Bibl. Anat.* **7,** 106–108.

Silva, J., and Intaglietta, M. (1974). The correlation of photometric signals derived from *in vivo* red blood cell flow in microvessels. *Microvasc. Res.* **7,** 156–169.

Starr, M. C., and Frasher, W. G. (1975a). A method for the simultaneous determination of plasma and cellular velocities in the microvasculature. *Microvasc. Res.* **10,** 95–101.

Starr, M. C., and Frasher, W. G. (1975b). *In vivo* cellular and plasma velocities in microvessels of the cat mesentery. *Microvasc. Res.* **10,** 102–106.

Wayland, H. (1973). Photosensor methods of flow measurement in the microcirculation. *Microvasc. Res.* **5,** 336–350.

Wayland, H., and Johnson, P. C. (1966). Erythrocyte velocity measurement in microvessels by a correlation method. *Bibl. Anat.* **9,** 160–163.

Wayland, H., and Johnson, P. C. (1967). Erythrocyte velocity measurement in microvessels by a two-slit photometric method. *J. Appl. Physiol.* **22,** 333–337.

Weber, E. H. (1838). *Arch. Anat. Physiol.* pp. 450–468.

Wiederhielm, C. A., Woodbury, J. W., Kirk, S., and Rushmer, R. F. (1964). Pulsatile pressures in the microcirculation of frog's mesentery. *Am. J. Physiol.* **207,** 173–176.

Wunderlich, P., and Schnermann, J. (1969). Continuous recording of hydrostatic pressure in renal tubules and blood capillaries by use of a new pressure transducer. *Pfluegers Arch.* **313,** 89–94.

8

Hemodynamics of the Microcirculation

I. THE CONCEPTUAL FRAMEWORK

From the viewpoint of fulfilling the purpose of the cardiovascular system, the area of most importance is the microcirculation. It is here that the heroic efforts of the heart, the large distributing arteries, and the collecting veins come to fruition. Broadly stated, this area makes it possible to optimally meet the variable needs of the dependent cells, tissues, and organs. The process of local homeostasis is dependent on the specific properties of the exchange vessels and their physical/chemical environment. In turn, each of these factors is dependent on microcirculatory hemodynamics. The principal physiological function of the microcirculation is inextricably linked to the magnitude and character of blood pressure and flow within the microvasculature.

Historically the physiologist, anatomist, research physician, and the experimentalist have, through the years, compiled a significant list of hemodynamic phenomena based on direct microcirculatory observation. Recently, biophysicists, bioengineers, and mathematically inclined researchers have addressed themselves to the issue of clarifying, characterizing and speculating upon many of the unique properties characteristic of the microcirculation. This task has been aided by the development of sophisticated computer capabilities for the purpose of numerical analysis, modeling, and simulation processes. In addition, the nature of the experimental data obtained in the recent

8. Hemodynamics of the Microcirculation

past has been sufficiently quantitative to provide concrete inputs to the theoreticians' models and analyses.

One of the important goals of microcirculatory research is to describe the pressure–flow distribution within the microvascular beds and to delineate clearly the relationship between these distributions and the significant factors which determine them. The description of these distributions (pressure and flow) have been obtained experimentally in certain microvascular preparations, but the full interpretation of this information requires the use of some form of conceptual model of the microvasculature. One such model is shown in Fig. 8.1.

The branching pattern illustrated in this figure is called dichotomous, in that each parent vessel gives rise to two offspring, each of the offspring give rise to two further offspring, and so on. This model of the microvasculature is useful to illustrate some conceptual points concerning the generalized terminal vascular bed distribution. However, as will be pointed out later, it cannot be taken too literally since, in fact, it is one of the extreme possibilities representing any real microvascular bed.

The dynamics of blood flow through the dichotomously branching system is rather easy to understand, which is one of the reasons for its usefulness. Blood flow (symbolized by the letter Q) is assumed to be flowing in the entering artery with unit magnitude. As shown by the arrow, the blood flow proceeds to the first bifurcation (also called branch point and node). At this site, by virtue of the complete symmetry of the vascular network, the blood flow divides equally, with one-half flowing to the left and one-half flowing to the right. This dichotomizing of the blood flow distribution continues at each successive bifurcation throughout

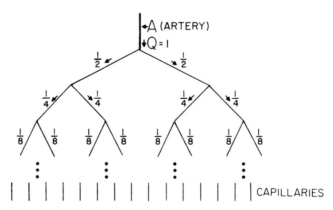

Fig. 8.1. Microvascular arterial distribution conceptualized as a single dichotomous branching system. The assumed complete symmetry requires all pressures at corresponding branch points to be equal and flows at each bifurcation to be half of the parent flow. The representation is idealized and departs significantly from the complex topography found *in vivo*.

I. The Conceptual Framework

the arterial distribution, finally reaching the arterioles and subsequently the capillaries. Though not illustrated in this figure, the venous collecting system would have an inverted confluent system, which would be the mirror image of the arterial divergent distribution system.

The highly idealized dichotomous model can be utilized to conceptualize the factors that affect pressure distribution throughout the microvascular bed.

To develop this concept, use will be made of the model shown in Fig. 8.2. For convenience, an unspecified microvascular bed is represented as consisting of seven vessel levels (sometimes referred to as vessel orders or orders of branching). The set of these vessels constitutes the pathway through which blood supplied from the aorta must flow to be collected in the vena cava. The simplest possible nomenclature for characterizing these seven levels has been applied progressively from large artery to large vein. Each of the segments of the microvasculature lying between consecutive dashed lines in the figure are associated with a specific resistance to blood flow. These resistances are designated as R_1 through R_7. For the topology of this conceptualized model, the resistance to blood flow offered by each consecutive segment depends on the resistance of each of the vessels within that segment. If sufficient data concerning the geometry, dimensions, and governing pressure–flow relationships within each individual vessel were available, these resistances could be calculated; however, such information is generally not available. The resistance of the first segment (the large artery) is simply the resistance of the large artery, itself, R_{LA}. Similarly, there is some resistance R_{SA} associated with each small artery. However,

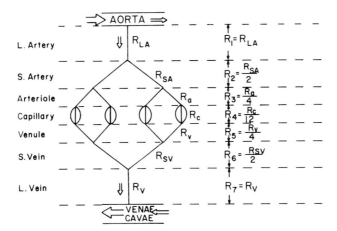

Fig. 8.2. Idealized representation of a microvascular bed as a series combination of different vascular levels. Each level is composed of different vessel designations, each of which has its own intrinsic resistance to blood flow. The set of vessels associated with each level determines the resistance to blood flow of that level.

since the two dichotomous branches at this level of the vasculature are effectively in parallel with each other, the net effective resistance of this level, R_2, is half of the resistance of each vessel. This analytical process can be repeatedly applied to each of the levels with the result that the resistance to blood flow offered by each of the seven levels can be expressed in terms of the resistance offered by each individual vessel. It is this resistance distribution (R_1 through R_7) that determines the pressure distribution. To follow this point, reference is made to Fig. 8.3; here, each of the consecutive levels of the microvasculature of Fig. 8.2 is represented as an equivalent resistance (R_1 through R_7), and each level is in series with each other. For convenience, we have taken the total resistance of the vascular bed to equal 100 units and have partitioned the resistance across each segment by an amount which is representative of the contribution to total resistance of each of the consecutive segments. Thus, the large artery segment R_1 contributes 1% to the total resistance; R_2, the small artery contribution, is 9%, and so on, such that the total of the individual contributions sums to 100%. In addition, the input pressure to the large artery is 100 mm Hg, a value not significantly dissimilar to the mean value found in this region of the circulation. The vena cava pressure is 0 mm

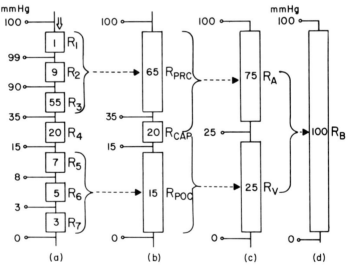

Fig. 8.3. Representation of microvascular bed resistances with different degrees of detail. Boxes represent vascular resistance of consecutive levels. Numbers in boxes are representative of the percent of the total resistance attributable to a particular level or combination of levels. Numbers outside of boxes show the pressure distribution for the assumed resistance profile. (a) All vessel levels of Fig. 2 shown separately. (b) Pre- and postcapillary resistances combined but capillary integrity maintained. (c) Vascular bed shown as pre- and postcapillary resistances. (d) Vascular bed viewed as a single lumped value.

I. The Conceptual Framework

Hg; hence, the total pressure difference from aorta to vena cava is 100 mm Hg. As can be seen from the figure, the pressure diminution from large artery to large vein is not linear but is reduced by an amount that is proportional to the relative resistance of each of the consecutive segments. Since the large artery contributes 1% of the total resistance, the pressure loss across the large artery segment causes a decrease in pressure from 100 to 99 mm Hg; the arterial level R_3, which represents 55% of the total resistance in this example, causes the pressure across this segment of the vasculature to be reduced from 90 to 35 mm Hg. It must be emphasized that the pressure at any level of the vasculature is determined by the ratio of the total distal resistance from that point to the total vascular resistance of the entire bed. For example, a pressure of 90 mm Hg at the entrance to the third level is determined from the fact that the sum of the distal resistances, namely R_3, R_4, R_5, R_6, and R_7 represent 90% of the total vascular resistance and hence 90% of the total pressure across the bed (i.e., 90% of 100 mm Hg is equal to 90 mm Hg). If it were possible to obtain specific and sufficiently detailed information regarding the resistance distribution throughout the vascular bed, it would be possible to approximate the pressure at any level of the vasculature simply by performing this ratio.

Because this specific information is lacking, it is convenient to divide the vasculature somewhat differently. Fig. 8.3 shows that all vascular resistance upstream or proximal to the capillary can be thought of as a single unit and defined as the precapillary resistance. Similarly, all the resistance downstream from the capillaries can be defined as the postcapillary resistance. Under some conditions, it is more useful to refer only to an arterial and a venous resistance. In this case, one-half of the capillary resistance is associated with the arterial resistance R_A, as shown in Fig. 8.3c, and the other half of the capillary resistance is associated with the venous side. Thus, a two-element model is produced as the vasculature is divided up into an arterial resistance, a venous resistance, and a pressure in between, the last of which reflects what is known as the midcapillary pressure. In the case illustrated, it equals 25 mm Hg. Finally, if it is of interest to characterize the total resistance of a particular vascular bed independent of how that resistance and/or pressure is distributed, then it is possible to lump all resistances together to define a single total value. This is illustrated in Fig. 8.3d. Each of the models illustrated in Fig. 8.3 has a particular utility which depends on its intended purpose. Thus, if the concern is with the resistance and pressure distribution of the microvascular bed then Fig. 8.3a would be an appropriate model. If the concern is mainly with events at the capillary level that are influenced by changes in pre- and postcapillary resistance, then conceptually either Fig. 8.3b or 3c may be used. Finally, the lumped characterization of the vascular bed, as illustrated in Fig. 8.3d, is useful when considering the total change in vascular resistance of the bed relative to, for example, changes in other vascular beds in the same animal.

II. MICROCIRCULATORY PRESSURE AND FLOW *IN VIVO*

Though the first microvascular blood pressure measurements were made over a century ago (Roy and Brown, 1879), it is only within the past 15 years that systematic *in vivo* data have begun to emerge. Until recently, specific information concerning the manner in which the blood pressure is distributed within the microvasculature and the relationship between microvascular and systemic blood pressure was lacking. Much of the insight into these and similar areas has been derived from direct pressure measurements in a few microvascular preparations. One of the most comprehensive of these studies is attributable to Zweifach (1974a,b). Using the cat mesentery as an experimental model and armed with the dual servo null micropressure system (see Chapter 7) almost 1000 pressure measurements in 110 animals were made. Interpretation of these extensive data proved to be no simple task. Though the conceptual representation of the microvasculature in terms of subunits described in Section I of this chapter provides a framework for this interpretation, the *in vivo* pressure measurements are made in an extremely complex vessel topographical arrangement. The vessel branching system is a mixture of dichotomous branching in certain areas and side branching in others. These patterns and their various permutations are intermixed and make the identification of branching orders, specification of vessel types, and development of conceptualization almost impossible. Zweifach's approach to this almost universal problem was to use a combination of vessel diameter and vessel position within the vascular network to characterize his results. Placing the capillaries at the center of a frame of reference, he could then specify vessel groups both proximally and distally. Arteriolar vessels, which most directly supply the capillaries, are termed precapillaries; those vessels immediately distal to the capillaries are termed postcapillaries. The precapillaries take their origin mostly as side branches from arterioles, and the postcapillaries drain into the larger muscular collecting venules. Thus, the conceptualization consisted of five vessel levels. However, due, in part, to wide variation in diameters of vessels which were referred to as arterioles (20–50 μm), Zweifach sought to make a finer distinction as applicable to the mesentery and defined the final 200–300 μm length of the arterioles to be the terminal arterioles. In addition to the comprehensive data provided by these studies, information on the pressures in the larger diameter arteriolar vessels has been provided by Gore (1974).

The conclusions concerning the microvascular pressure profile provided by these kinds of studies depends both on the particular tissue in which the measurements are made as well as the manner in which the data are assimilated. For example, when pressure distribution data are presented as an explicit function of vessel size groupings as seen in Fig. 8.4, data seem to confirm the fact that consecutive levels of the microvasculature do not produce equal reductions in pressure. As might be expected, the change in pressure with respect to microvas-

II. Microcirculatory Pressure and Flow *In Vivo*

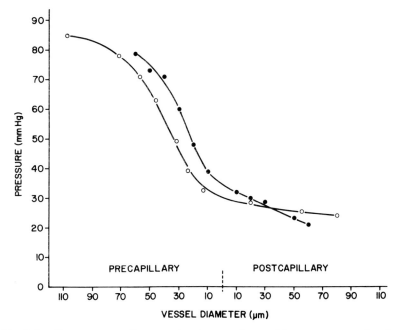

Fig. 8.4. The pressure profile across the microvasculature of the cat mesentery plotted as a function of vessel diameter groupings. All data points represent mean values. Data: ●, Zweifach (1974a); ○, Gore (1974).

cular level is significantly greater on the arterial side of the capillary than on the venous side, where an almost linear relationship is found. Contrastingly, on the arterial side, the pressure dependence on microvascular level appears nonlinear and the largest pressure drop is stated to be produced by the terminal arterioles and precapillaries. This basic dependence of microvascular pressure on vascular level and associated vessel diameters is consistent with less extensive measurements in the mesentery (Gore, 1974), intestinal skeletal muscle (Bohlen and Gore, 1977), and in the cremaster musculature (Bohlen *et al.*, 1979). However, as can be seen in Fig. 8.5, significant differences in the details of these pressure profiles are evident. Indeed, pressure profile characteristics as obtained in other tissue preparations, which include the cat tenuissimus muscle (Fronek and Zweifach, 1977), cat omentum (Intaglietta *et al.*, 1971), rabbit omentum (Gross *et al.*, 1974), and bat wing (Nichol, 1969; Wiederhielm and Weston, 1973), all differ in detail.

The pressure profiles drawn in Fig. 8.5 appear to show a significant nonlinearity characterized by a rate of pressure loss highest in the range of the smaller arteriolar vessels. However, when the same data are used to draw the pressure profile as a function of microvessel level, a very different picture emerges (Fig.

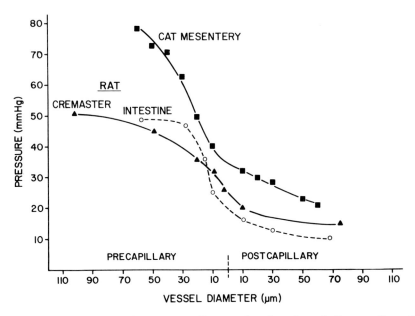

Fig. 8.5. Comparison of pressure profiles as a function of vessel diameter. Data: ○, (Zweifach, 1974a), cat mesentery; ■, Bohlen and Gore (1977), rat intestinal muscle; ▲, Bohlen et al. (1977), cremaster muscle of spontaneously hypertensive rat.

8.6). It now appears that microvascular pressure loss is an almost linear function of vascular level, with arterial and venous regions simply characterized by different slopes. The obvious implications of this kind of property (if borne out by further experimental work) is that under certain circumstances, the relative resistance to blood flow is uniformly distributed throughout the microvascular bed. Under such conditions it would no longer make sense to talk about "a principal site of arterial resistance" classically attributed to the loosely used term, arteriole. A likely possibility is that, depending on a variety of as yet undetermined parameters, the microvascular pressure profile is in fact a variable which may undergo considerable change. Evidence that strong neural excitation can modify pressure significantly has just recently become available (Bohlen and Gore, 1979).

Once the characteristics of a microvascular pressure profile are determined under conditions controlled so that proper *in vivo* measurements can be made, a natural area of inquiry is the dependence of microvascular pressure on systemic blood pressure. Deviations in systemic blood pressure from "normal" values occur naturally due to statistical variations among individuals. Other changes are associated with exercise and stressful situations. Pressure is elevated in a variety of forms of hypertension and is reduced in hemorrhage, to cite only a few

II. Microcirculatory Pressure and Flow In Vivo

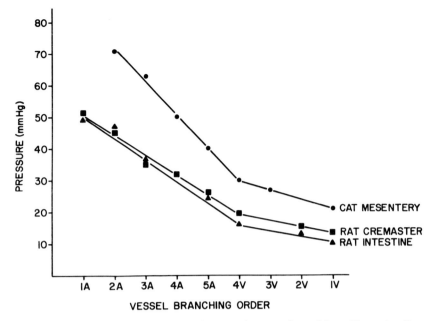

Fig. 8.6. Comparison of pressure profiles as a function of vessel branching order. Data: Same as Fig. 8.5. A, precapillary; V, postcapillary. Profile suggests a linear change in pressure with microvascular level.

examples. What, under these conditions happens in the microcirculation? Information on this important issue is both scarce as well as controversial. Zweifach, in analyzing naturally occurring systemic blood pressure variations, found a moderate correlation between the systemic blood pressure and the microvascular blood pressure only in vessels greater than 50 μm. Conversely, Gore (1974), who controlled systemic pressure, reported a linear dependence between systemic pressure for arteriole vessels as small as 24 μm, and later Bohlen and Gore, (1979) reported a linear dependence of capillary pressure on systemic pressure independent of innervation. Data from hypertensive animals suggests that the fraction of the systemic pressure at all microvascular levels is the same as within normal animals. What could account for such behavior? A linear dependence of microvascular pressure on systemic pressure at all vascular levels can occur in one of two ways; (1) the entire vasculature is behaving as a multitude of interconnected rigid pipes, or (2) the sum effect of all vascular control systems acts to produce a microvascular resistance distribution which is independent of systemic pressure. The first possibility seems ridiculous and the second seems unlikely. Further work in this important area is required to help resolve the problem.

One final note concerning pressure distribution in the microvasculature is in order. Of necessity in all experimental studies designed to establish the normal distribution of pressure within the microvasculature, use has been made of values computed as the average for a particular vessel size or branching order. A perplexing finding is that pressure magnitudes at apparently the same microvascular level within vessels of the same diameter range display a tremendous spread. Such variability in the measured quantities undoubtedly reflects the normal vasomotion of vessels proximal and distal to the site of measurement. However, since the bulk of the data strongly suggest that the dispersion of pressure is least in the capillary region, it is clear that the vessel vasomotion which causes temporal variation in distal pressures cannot completely account for wide differences found in different vessels in the same animal or in different animals. It is probable that the specific site within a particular vessel grouping accounts for some of this pressure variability, which must also reflect how close one is to the proximal or distal end of the vessel. An additional important factor that probably gives rise to this variability is the complex heterogeneity of the vascular topography in which the measurements are made. It is therefore imperative that suitable time and attention be taken to carefully describe and characterize in sufficient detail the topography of the vasculature in which all such micropressure measurements are made.

Pressure and its distribution in the microvasculature allows one to determine the *relative* resistance associated with each level of the vasculature. Comparison of control pressure profiles with those obtained subsequent to experimental interventions or accompanying pathological states permit one to assess changes in this distribution and thus gain insight into hemodynamic derangements. However, from both physiologic and hemodynamic viewpoints, it is of equal importance to establish the character of microcirculatory blood flow patterns and distributions. The principal method currently used for this purpose is based on the dual-slit photometric technique perfected by Wayland and Johnson (1967) (see Chapter 7). Blood velocity is the directly determined quantity from which volumetric flow is calculated using vessel diameter and appropriate empirical correction factors.

Whether one is interested in determining *in vivo* blood velocity in individual vessels or in characterizing the velocity or flow distribution throughout the microvasculature, certain fundamental principles should first be considered. For this purpose, use will be made of Fig. 8.7, which depicts a segment of an arterial vessel and two equal length branches which have uniform but nonequal diameters. The resistance of each branch segment, R_{b1} and R_{b2}, will differ by an amount inversely proportional to the fourth power of their respective diameters. Since the details of the network topology distal to the level $X—X$ are not of interest at present, each intricate pathway is simply replaced with lumped quantities which represent the total vascular resistance between $X—X$ and the vena

II. Microcirculatory Pressure and Flow *In Vivo*

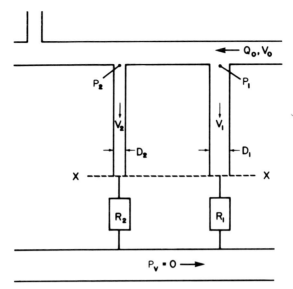

Fig. 8.7. Illustration of the fact that values obtained from measurements of local erythrocyte velocity depend on both local and global conditions.

cava. The point to be realized is that the locally measured values of V_1 and V_2 depend not just on local conditions, but on the overall distal resistance as well! Further, it is even possible under appropriate conditions for the velocity measured in an individual branch to be *larger* than that in its parent vessel, though volume flow must always be less. To amplify this point, it should be noted that the measured velocity is proportional to the ratio of the volume flow (Q_i) to the square of the vessel diameter (D_i^2) (Eq. 1).

$$V_i = K \frac{Q_i}{D_i^2} \qquad i = 1, 2 \qquad (1)$$

Since the branch flow itself depends on the ratio of the pressure difference to total branch pathway resistance, the measured velocity can be seen to depend on local and global conditions.

$$V_i = K \frac{P_i}{D_i^2} \frac{1}{R_i + R_{bi}} \qquad (2)$$

Depending on the relative values of global versus local quantities, V_2 can be less than, equal to, or greater than V_0 and/or V_1.

Though such parent–branch velocity relationships described above do occur *in vivo*, the average trend for the velocity profile across the microvasculature is one characterized by a decrease of blood velocity with branching order. This has been established for all tissue studies in a systematic fashion and is illustrated in Fig. 8.8 for two quite different tissues; the bat wing (Mayrovitz *et al.*, 1977) and cerebral microvessels (Ma *et al.*, 1974). Since these tissues differ dramatically both in function as well as weight, the magnitude of the absolute difference in velocity is not surprising. However, it should be noted that as the capillary level is approached, the velocities approach each other. Thus, it appears that not only is pressure in the capillary region subject to much less variation across species in animals but this is also true for red blood cell velocity. The velocity profiles for most tissues investigated to date, including skeletal muscle (Fronek and Zweifach, 1977) and cat mesentery (Lipowsky and Zweifach, 1977), fall within the bounds described by the two curves of Fig. 8.5.

The advent of instrumentation capable of tracking rapid changes in blood velocity (and pressure) put the study of the pulsatile nature of microvascular hemodynamics on firm ground. In 1971, using the cat omentum, Intaglietta *et al.*(1971) employed simultaneous measurements of microvascular pressure and

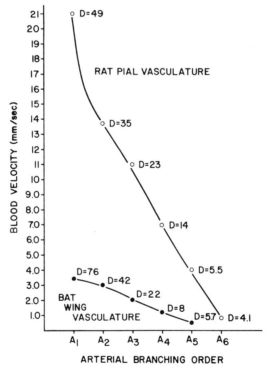

Fig. 8.8. Illustration of the velocity profile across the microvasculature of two quite different tissues. Data: ○, Ma *et al.* (1974), rat pial vessels; ●, Mayrovitz *et al.* (1977), batwing vessels. D, mean vessel diameter (μm) encompassing each vessel order.

II. Microcirculatory Pressure and Flow *In Vivo*

velocity to establish that significant pulsatile components of each are present. The magnitude of this pulsatile velocity component and its dependence on the branching level has been systematically explored in the bat wing, and a typical pattern is illustrated in Fig. 8.9. As may be seen, the effect of the branching order on the microvascular level is to reduce both the mean and peak components of blood velocity. Interestingly, however, calculations reveal that even at the capillary level, the ratio of pulsatile to mean velocity is about 0.25, which indicates a significant pulsatile component. Analysis and interpretation of such microcirculatory velocity as well as pressure data are in some ways more, and in other ways less complicated than with pulsatile events in large distributing arteries. Included among factors that render analysis more difficult are the multiplicity of blood vessels, the extensive and nonuniform branching pattern, the difference in vessel geometry and wall properties associated with different branching levels, and the nonuniformity of erythrocyte concentration and hence effective viscosity. Partially offsetting these difficulties is the fact that calculations show the hemodynamics within the microvasculature to be governed by values of Womersley's alpha that are much less than unity and by wave propagation speeds sufficiently small so that vessel lengths are small in comparison with the pulsatile

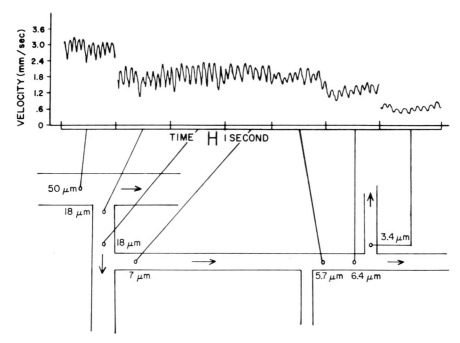

Fig. 8.9. Pulsatility of blood velocity synchronous with heart rate present in the microcirculation of the bat wing. Data: Mayrovitz *et al.* (1976b).

wavelengths. Thus, for many purposes pulsatility of the microcirculation can appropriately be analyzed by neglecting intertial effects and the distributed nature of wave propagation. However, the significance both hemodynamically and physiologically, of the presence of pulsatile components of both pressure and flow in the microcirculation is almost as much a mystery today as it was one-hundred years ago.

In addition to hemodynamic variations synchronous with the heart rate, an extensive temporal variability in microcirculatory pressure and flow is a commonly observed phenomenon. One major cause of this variability is due to active and dynamic diameter changes of precapillary and postcapillary microvessels serving a local control function and collectively referred to as vasomotion.

The spontaneous diameter changes associated with the terminal arterioles and precapillary sphincters produce a modulation of the pressure and flow in the capillary networks. This modulation has several effects. The decrease in diameter is associated with a drop in pressure at the arterial end of the capillary by an amount which depends on the amount of diameter change. For the diameter changes most frequently observed experimentally, the capillary pressure is reduced from a value of about 40 mm Hg to a value close to the venous pressure (Mayrovitz et al., 1978). This may mean that the contraction phase of vasomotion reduces capillary pressure to a value below the commonly accepted normal value of plasma osmotic pressure, and it implies that the contraction phase of vasomotion is associated with capillary reabsorption over the entire capillary length for normal arterial pressures.

Study of the postcapillary microvasculature (Mayrovitz, 1974) has yielded evidence indicating that, in addition to the precapillary dynamics, the rhythmic diameter variation of muscular venules and small veins plays a significant role in determining the flow and pressure distribution in the microcirculation. Though the fundamental mechanisms controlling these dynamics have not yet been fully elucidated, the responses of these postcapillary vessels to pressure perturbations and the presence of vascular smooth muscle in the walls of these vessels clearly suggest that there is a myogenic component involved. The persistence of these dynamic diameter changes in the denervated preparation further supports the idea of a local mechanism associated with active contraction. Detailed examination of the characteristic properties of this process has shown that elevations in frequency secondary to pressure elevation stimuli are produced by an increase in the rate of distention, with no significant change in the time course of the contraction phase. This contraction is associated with an increase in segmental vascular resistance and vessel wall thickness and a decrease in vascular compliance. The contraction causes the instantaneous pressure within these vessels to rise by an amount which depends on the absolute value of the venous pressure existing prior to the contraction; the pressure rise ranges from 6 to 10 mm Hg. These values represent a significant increase in the venule and small vein pressure in view of the normal mean pressure levels in these postcapillary vessels (10–15 mm Hg).

The amount of control exercised by postcapillary dynamics depends on the state of the terminal arteriole and/or the precapillary sphincter supplying the dependent capillary region.

The most dramatic effects on capillary flow are related to alterations in the precapillary sphincter state. However, when the arteriolar vessels are in a contracted state, the modulation in capillary flow due to post capillary vasomotion may be 50% or larger. This would imply that under normal conditions the significance of post capillary dynamics on capillary flow modulation waxes and wanes with alternate contraction and dilation phases of arteriole vasomotion. From this it follows that increases in relative contraction time of the sphincter or terminal arteriole would assign to the postcapillary vasculature and associated dynamics an increasingly significant role as determinants of capillary flow distribution and as factors in venous circulation. One may speculate, then, that physiological conditions which tend to augment the constrictor phase of arteriole vasomotion will simultaneously increase both the local and systemic importance of the postcapillary vasculature. One such instance satisfying this condition would be systemic hypertension.

III. THE SIGNIFICANCE OF DESIGN, CELLS, AND CAPILLARY FLOW

A. Vascular Design

Concern about the significance of vascular design has a moderate but somewhat tortuous history. The general areas of inquiry can be divided into two broad categorizations. The first was concerned with the concept of optimality, introduced by Murray in 1926 and built upon by others (Cohn, 1954, 1955; Zamir, 1977). This approach is an investigative path in which likely vascular parameters (vessel diameter, number of branches, lengths of vessels, angles of branching, and the overall topography of the vasculature) are chosen to be optimized in some functional sense, and the vascular configurations that will satisfy these imposed optimality conditions are synthesized, analyzed, and compared to reality. The results so far of this approach have produced configurations which, though idealized, in most cases do compare in some aspects to what is observed under the microscope. This kind of approach, aimed at clarifying the complex nature of the vasculature and its significance, is still used today.

The second approach to studying vascular design, which will be dealt with in greatest detail, is to use what is known about topography and metrics of actual vascular beds to create a simplified representative model of the microvascular bed that still includes essential details necessary for the type of analyses desired (see Fig. 8.10).

The representation of Fig. 8.10 derived from the structure of the bat wing

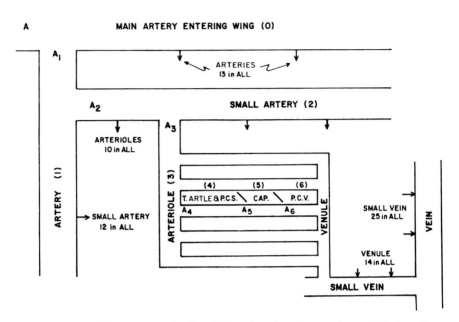

Fig. 8.10. Idealized topographical model based on the microvasculature of the bat wing. (From Mayrovitz et al., 1976.)

vasculature is one of the first attempts to apply this technique (Mayrovitz et al., 1975). Presently, there are other representations that are applicable to different preparations, a fact that must not be lost sight of. Once the structural characteristics of the vascular bed are established, it is necessary to characterize, in a mathematical sense, the relationships as well as the interrelationships between the vascular structure and important hemodynamic variables. With reference to Fig. 8.11, one approach to such a characterization is to represent a single vessel of any branching order in terms of a series component; thus, r_i is designated as the resistance to blood flow offered by each of the segments of the vessel, and R_{i+1} is the resistance of the branches from the site at which they emerge from the parent vessel as shunt components. This is quite a general representation of vessel structure and topography for such a complex branching system. Most representation of this type have of necessity utilized average data from which resistances have been calculated. Once the structure has been established and the pertinent hemodynamic equations written to describe the pressure–flow relationships in each of the various levels, it can then be utilized to compare calculated results and predictions with those obtained from experimental data. One such comparison is indicated in Fig. 8.12, where the pressure distribution across the

III. The Significance of Design, Cells, and Capillary Flow

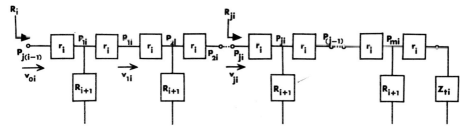

Fig. 8.11. Detailed representation of each of the vessels shown in Fig. 8.10. Each vessel is represented by as many "T" sections as there are branches. This figure illustrates the general relationships for a small ith order vessel; for example, if $i = 2$, r_i represents the longitudinal resistance of the small artery (second order branch) and R_{i+1} the total resistance looking toward the capillary at each of the branching sites of the small artery, i.e., looking into the arteriole. The pressure at the origin $P_{j(i-1)}$, is the pressure at the jth branch of the $(i-1)$ vessel (artery), and the pressure P_{ji} serves as the input pressure for the $(i+1)$ order vessels. The quantities R_{ji}, P_{ji}, and v_{ji} denote mean resistance, pressure, and velocity at the jth segment of the ith order vessel. The quantity R_i is the total resistance seen looking into the ith order vessel at its origin. (From Mayrovitz et al., 1976a.)

vascular bed is compared with predictions based on the analytical model described by Figs. 8.10 and 8.11. Various models of this type have had reasonable success in predicting overall vascular bed hemodynamics and have provided general insight into important factors influencing microvascular hemodynamics. Extensive measurements of lengths and diameters in the microvasculature of the rabbit omentum (Intaglietta and Zweifach, 1971), supplemented by pressure and flow measurements, led to the development of a simplified electrical equivalent model of the microvasculature that includes elements of vascular resistance and compliance (Intaglietta et al., 1971). Morphologic and structural properties were used by Gross and Intaglietta (1973) to model and calculate the resistance, compliance, and pulsatile pressure distribution across the microvascular bed of several different species. The modular configuration of the microcirculatory system of the cat mesentery was subjected to a network analysis approach (Lipowsky and Zweifach, 1974) to determine the microvascular pressure profile and compare it with available *in vivo* data. Significant differences were demonstrated, especially with regard to the gradients of pressure profile and the pressure gradient in small arteriolar vessels. The bat wing vasculature proved to be an ideal representative microvascular bed from which microvascular models were developed to study the effects of vessel dimensional changes on microvascular hemodynamics and the role of a variety of rheological complications unique to the microcirculation (Mayrovitz et al., 1975, 1976). The complex interactions between hemodynamics, topology, vasomotion, and rheology within the microcirculation was then attacked using an extensive but representative model (Mayrovitz et al., 1978). Results from these models provided insight into many

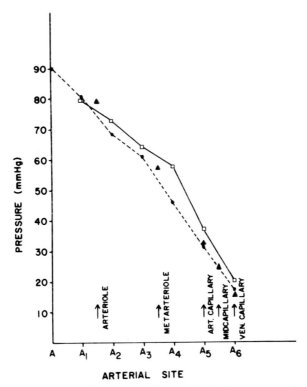

Fig. 8.12. A comparison of the pressure distribution in the microvasculature of the bat wing as predicted from model analysis and that measured *in vivo*. ──□──, calculated from the model; --●--, calculated from velocity distribution; ▲, measured. (From Mayrovitz *et al.*, 1976.)

dynamic aspects of the microcirculation and offer predictions of hemodynamic behavior associated with pre- and postcapillary vasomotion that are inaccessible to *in vivo* measurement. In spite of the inroads made into the general hemodynamic properties provided by this combination of experimentation, model development, and analysis, little systematic effort has been devoted to the very terminal microvasculature, i.e., the arterial bed–capillary interface and the possibility of variations of flow within this region.

As has been noted in previous chapters, one of the parameters used to estimate the status of the capillary circulation is the concept of capillary density, which can be expressed either as the number of capillaries observed per unit length, per unit area, or extrapolated to per unit volume. In Fig. 8.13, four sets of configurations are schematized; each has a uniform linear capillary density. It should be noted that the most significant parameter of capillary density is probably the number of capillaries per unit volume of tissue. However, a three-dimensional

III. The Significance of Design, Cells, and Capillary Flow 195

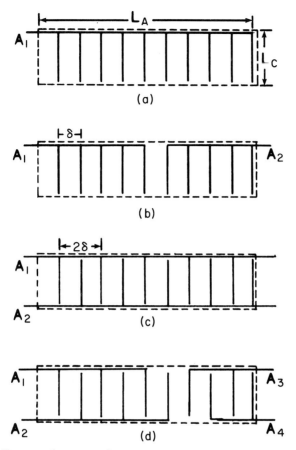

Fig. 8.13. Four vascular topographies each of which yields the same capillary density. A_1 through A_4 denote arterioles supplying capillaries of length L_c.

representation, though offering more degrees of freedom, is much more difficult to represent. It is felt that the basic elements of the planar case still embody the important basic concepts. In Fig. 8.13, the dashed area represents a region of tissue having a certain length, width, and surface area. In each of the four cases illustrated, the condition of a uniform capillary density is imposed by assuming that each capillary is separated from its neighbor by a uniform distance δ. Each of the configurations illustrated satisfies the condition of uniform and equal capillary densities in a different fashion. In Fig. 8.13a, M capillaries are subserved by a single arteriole. In Fig. 8.13b, the same tissue region is supplied with the same capillary density, but now as a consequence of branches of two arterioles. Note that in Fig. 8.13b, each of the two arterioles has an equal number of branches. This is not necessarily a requirement to satisfy the condition of equal capillary density. Thus it would be possible, for example, to lengthen the arteriole denoted as A_1 and to shorten the arteriole denoted as A_2 while still maintaining the same capillary density simply by reducing the number of branches from arteriole A_2, and by increasing the number of branches from arteriole A_1. Such an arteriole distribution is completely consistent with physiological observations. Figure 8.13c illustrates the situation in which the same capillary density is maintained but by a different arteriole distribution. Indeed, as in Fig. 8.13d, there are still two arterioles supplying the dependent tissue region, but the pattern of their network is different in that the location of the branches of each supplying arteriole is separated by twice the distance previously shown. The way in which a uniform and equal capillary density is achieved is to have an intermeshing of these capillaries so that the effective linear capillary density is given by the capillary spacing δ. Figure 8.13d shows an arteriole distribution consisting of four arterioles of the appropriate topography to produce a tissue capillary density exactly equal to that of the other configurations. From a hemodynamic point of view, the four configurations illustrated in Fig. 8.13 are different, even though they achieve the same capillary density. Each representation will have its own appropriateness, depending on the particular vascular tissue under consideration as well as on the region within that vasculature. However, it is possible to characterize some of the general properties that pertain to all of them and to provide a framework for those who are exploring tissue preparations in which one or more of the configurations is applicable.

In Fig. 8.14, an arteriole and its branches (capillaries) are conceptualized as being represented by their respective hemodynamic impedances. The term impedance is more general than resistance and includes, in addition to viscous losses, the possibility of inertial and compliant effects. Z_1 represents the impedance of the parent arteriole lying between branches, and Z_2 represents the total impedance seen at the point of branching looking toward the capillary. In addition, the end of the arteriole is terminated in an impedance Z_L, which represents the total impedance seen at the final ramification of the distributing arteriole. It should be pointed out that this kind of conceptualization of a vessel and its branches would apply to any level of the vasculature as long as the topography of the vascular network was similar. One of the important hemodynamic parameters associated with any vascular network is the input impedance to blood flow which that vascular network offers. This is characterized in the model as Z_{in}. The importance of the input impedance in part revolves around the fact that the power consumed by the vascular network is directly proportional to the input impedance for the same total flow to the capillary network. All things being equal, a larger Z_{in} would necessitate a greater consumption of power in order to satisfy the tissue

III. The Significance of Design, Cells, and Capillary Flow

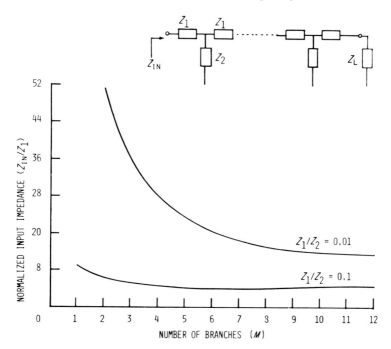

Fig. 8.14. Input impedance of a single uniform diameter vessel with multiple branches. For any given value of parent vessel segmental impedance (Z_1) the input impedance can be slightly or greatly dependent on the number of branches. The determinant factor is the branch impedance (Z_2).

needs. In addition, the blood flow is inversely related to the impedance. It is important to discover how the input impedance depends on the details of the vascular network. It turns out that the Z_{in} of this type of structure is highly dependent on the ratio of the impedance of the parent vessel to the impedance of the branch. For conciseness we shall refer to the parent as an arteriole and the branch as a capillary. If the ratio of arteriole to capillary impedance is, for example, 0.1 (capillary impedance is ten times that seen in the arteriole segment), then almost independently of how many branches that arteriole has, the input impedance (normalized to that of the parent vessel) does not change. However, if we contrast this with the situation in which the capillary impedance is 100 times that of the supplying arteriole, then there is a strong relationship between the number of branches that the arteriole has and the relative input impedance seen. One of the important consequences of this kind of behavior is that, depending on the detailed nature of the vessel diameters, lengths, and number of branches, very different hemodynamic behavior may be anticipated.

This requires that formulations of the appropriate theoretical structure be based on very careful experimental data and points out that specific measurements of the diameter are crucial to a proper hemodynamic analysis.

Figure 8.14 characterizes the arteriole–capillary distribution as being from one parent arteriole. It is fairly commonly observed that in many circumstances two or more arterioles supply a given tissue region. For the purpose of illustration (Fig. 8.15), consider how the input impedance of two arterioles supplying a specified tissue region depends on the number of branches (capillaries) of each of the arterioles. In effect, consider the topography shown in Figs. 8.13b and 8.13c to see what can be determined about the general behavior of such a vascular pattern. For each of the four curves in Fig. 8.15, the total number of branches from the two arterioles is constant, but the manner in which they branch from these two vessels is different. For the curve marked $K = 6$, there are a total of six capillaries emanating from two arterioles; for $K = 8$, eight capillaries, etc. The abscissa designates the number of branches from one of the arterioles (M_1) and hence the difference between K and M_1 would be the number of branches from the second arteriole. This procedure is extended to consider the cases of the total

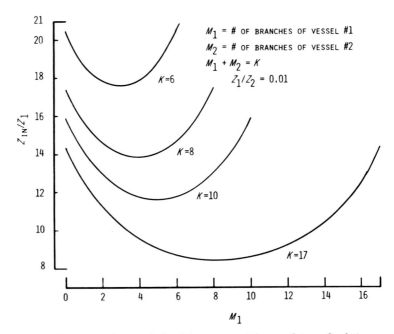

Fig. 8.15. Input impedance calculated for two arterioles supplying a fixed tissue region such as shown in Figs. 8.13b and c. An impedance minimum occurs when each vessel has an equal number of branches. The magnitude of this minimum value depends on the actual number of branches.

III. The Significance of Design, Cells, and Capillary Flow

number of branches equal to 8, 10, and 17. What is found out about the general properties from this kind of analysis? First, it is seen that for any given branching sequence, there is always a minimum in terms of the input impedance seen looking into that vessel. If there is any significance to the concept of minimization of power or work then a question arises. Does the architecture of the vasculature operate in a system whereby two vessels will develop in a manner so as to approach a minimum impedance situation? An important general observation is that as the length of the vessel increases, or as the number of branches increases, the position of that minimum moves both down and to the right. From this simple analytical procedure it may be concluded that the greater the number of branches present, the lower will be the input impedance. Secondly, for any given arteriole length, a minimization in input impedance can be achieved if the two arterioles have approximately the same number of capillary offshoots. This raises the important experimental question as to the details of how the terminal arterioles and the capillaries are precisely distributed—a question not yet answered.

At this point, a different complication is introduced, i.e., vessel taper. Vessel taper, with a few notable exceptions (Jeffords and Knisely, 1956; Skalak and Statitis, 1969; Walawender and Chen, 1975), has been ignored both experimentally and as a component of the analytical consideration of the terminal vascular bed. However, it is present; thus we ask how significant is the taper and what are its effects? Again it must be emphasized that the amount of taper depends upon the region of the microvasculature one is examining as well as the specific vascular bed being studied. To illustrate the kind of behavior to be expected from a tapering vessel, an arteriole that has a certain inside diameter at its origin (D_{in}) and a certain diameter at its exit (D_o) is specified in Fig. 8.16. As before, a specific tissue region (dashed lines) is defined which is supplied through the arteriole. The capillary density is denoted by a linear capillary separation δ, and the flow per capillary as Q_c. In order to be quite specific, an analysis is performed for the case in which the assumed taper is 2.5×10^{-3} microns per micron of arteriole length. In this manner the pressure distribution along the arteriole as well as the blood flow distribution into the capillaries can be studied.

Figure 8.17 shows how important the introduction of vessel taper may be. Each of the three curves illustrated in this figure are for different capillary spacings and thus different effective capillary densities. Two important concepts which emerge are that vessel taper will result in a minimum in input resistance at some specified vessel length, and the magnitude of that minimum is crucially dependent on the capillary density of the tissue it subserves. Generally speaking, the higher the capillary density, the lower is the input resistance for any corresponding length. From a functional point of view, it would be expected that an increase in capillary density would find a decrease in input resistance advantageous to supply the concomitant increased flow demand.

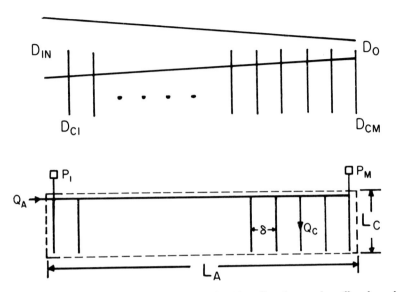

Fig. 8.16. Representation of a tapering arteriole with uniformly spaced capillary branches supplying a specified tissue region.

The input resistance depicted in Fig. 8.17 is only one part of the hemodynamic story associated with tapered vessels of variable length and/or variable capillary density. The total inflow to the arteriole is dependent on the input resistance, but also on the perfusion pressure. Further, given a certain input flow, one should inquire as to how that flow is distributed to the capillary network. To examine this conceptually it is possible to assume a certain input blood pressure (75 mm Hg) and hold that level constant. One can then choose a realistic capillary density (say one capillary/100 μm), impose the taper as was done for the analysis of Fig. 8.17, and calculate the total inflow and the flow in each capillary as the length of the arteriole is varied. The results of one such analysis are illustrated in Fig. 8.18. For the particular structure and diameters chosen for this example, a maximum in arteriole inflow is found to occur at the input resistance minimum. However, even though the total flow entering the arteriole is maximized under these conditions, the flow per capillary decreases monotomically as the arteriole length is increased. This result is obtained because, for a fixed capillary density, an increase of the actual length of the arteriole supplying that region must be accompanied by an increase in the total number of capillary branches. The surprising consequences of this kind of behavior (with or without vessel taper) is that it predicts that capillaries more distal (in terms of their point of origin from the entrance site of the arteriole) will receive less blood flow than those more proximal. Experimentally, there is no data that can offer a reconciliation of this question. It is possible that because of local vasomotion and other undefined

III. The Significance of Design, Cells, and Capillary Flow

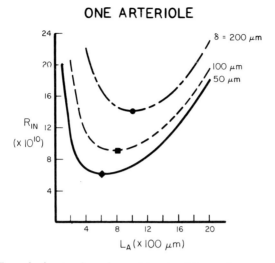

Fig. 8.17. Example showing dependence of input resistance of a tapered vessel on vessel length and capillary density. Parameters: $D_{in} = 11\mu m$, $D_o = 5.5\mu m$, $D_{cl} = D_{cm} = 5.5\mu m$. Note that for fixed vessel length the input resistance falls with increasing capillary density and that for each capillary density there is a unique arteriole length for which R_{in} is minimum.

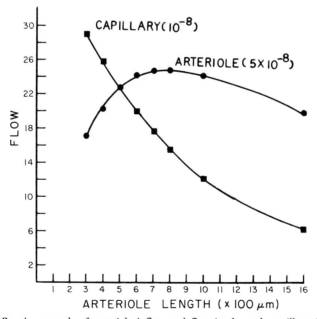

Fig. 8.18. An example of arteriole inflow and flow in the mth capillary for tapering arteriole–capillary model illustrated in Fig. 8.16. Note that for a fixed capillary density (1 capillary/100 μm) there is a peak in the arteriole inflow corresponding to a certain arteriole length but that the flow in the final capillary continues to decrease.

hemodynamic phenomena there is a more uniform perfusion state than would be forecast by this kind of analysis.

The question of whether or not capillaries more distal from the point of origin of their parent arteriole receive less flow is still open. However, it is worthwhile to explore this question to provide insight into factors which may influence capillary flow behavior. To accomplish this, two very different hemodynamic cases shall be contrasted. In the first, all capillaries from the parent arteriole have equal resistance to blood flow; in the other case, all capillaries that emanate from the same arteriole have equal capillary blood flow.

The conceptual diagram illustrating these two cases are shown in Fig. 8.19. In part a of this figure, an arteriole with the length of 1200 μm is shown giving rise to uniformly spaced capillaries of unspecified number. Each capillary has a length (l_c) equal to 500 μm and each, including the mth capillary, has an effective inside diameter of 6 μm. Since the lengths and diameters of each of the capillary branches are equal, then each capillary pathway will have the same resistance to blood flow. Parts b and c of this figure illustrate two cases in which the uniform resistance case can be transformed into a case of equal capillary flow. It is required that the diameters of each capillary more proximal to the origin of the arteriole have diameters that are less than the last capillary and, in fact, progressively less, as the origin of the arteriole is approached. To satisfy the equal flow condition, the diameter of the furthest capillary from the end need be only 1.5 μm less than the largest capillary, even if this arteriole gives rise to twelve capillaries. Capillaries in between the first and the last would have intermediate values. If there were a smaller number of capillaries between the most distal and most proximal capillary, then the diameter difference necessary to achieve a condition of equal capillary flow would be much less. Thus, it can be seen that very small differences in capillary dimensions, even approaching the resolution with which these diameters can be resolved *in vivo*, make significant differences with respect to the actual flow distribution. Again, no substantial *in vivo* information is available on this point.

Figure 8.13c shows another way in which an equal flow per capillary can be produced while maintaining essentially the same internal diameter of each capillary. The increase in capillary resistance, as the proximal end of the arteriole is approached, is achieved by an increase in the length of the capillaries more proximally located. The lengths and diameters chosen in this example are characteristic of those seen in physiological situations; the requirement for uniform flow per capillary is that the most proximal capillary length be about 1800 μm, or about three and one-half times the most distal capillary, even for as many as twelve capillaries per arteriole.

From an experimental point of view, no information is available that systematically evaluates the lengths of capillaries in relation to their position in the arterial distribution. Yet, the hemodynamic significance of such vascular designs are clearly evident.

III. The Significance of Design, Cells, and Capillary Flow

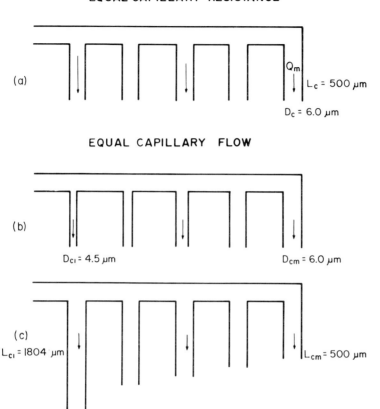

Fig. 8.19. Schematic representation of an arteriole–capillary distribution contrasting equal resistance and equal flow per capillary. (a) All capillaries have uniform lengths, diameters, and equal resistance to flow. (b) All capillaries have equal flow due to slight reduction in diameter progressing from the distal end. (c) All capillaries have equal flow due to progressive increase in their lengths. In each case arteriole length is $1200 \mu m$ and diameter is $9.0 \mu m$. Note that a small difference in capillary diameter can convert an equal resistance structure into one in which there is an equal flow per capillary.

Analysis of the pressure loss as a function of capillary position for the above cases as shown in Fig. 8.20 indicates that there is not much difference between having equal flow per capillary and equal resistance per capillary; however, in the latter case, the pressure loss per branch is slightly larger. Thus, it is probable that for the expected variations in pressure, *in vivo* pressure measurements would probably not be revealing in this regard. The large absolute value of pressure loss as a function of position may account for some of the wide variations and scatter reported in the experimental literature on pressure measurements in the microcirculation.

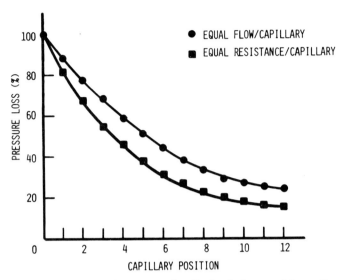

Fig. 8.20. The pressure loss from proximal to distal end of an arteriole–capillary distribution. The details of this distribution depend on dimensions and the number of capillary branches. Note that the equal flow/capillary case produces slightly less pressure loss, but in both cases the pressure at the distal capillary is significantly reduced.

Though the pressure distribution along the arteriole is insignificantly different, flow distribution is quite different, as shown in Fig. 8.21. The straight horizontal line defines this situation in which each capillary has a flow of 2×10^{-8} CGS units. This corresponds to the equal flow/capillary case. If the true nature of the microvasculature is more closely one of uniform resistance per capillary, but a minimum tissue flow requirement per capillary is present, then this property can be simulated by forcing the last capillary to receive that minimum value. When this is done, it turns out that capillaries which are proximal to the last one will always receive an excess flow to meet the imposed distal requirement. Again, however, it must be emphasized that local control mechanisms may significantly mediate this process.

The manner in which these local controls, as well as a variety of other flow-influencing phenomena, affect capillary blood flow is dependent on a variety of factors, including the precise detail of the terminal microvascular network topology. Temporal variability in capillary flow, described by Krogh (1929) and variously referred to as intermittent or waxing and waning, is a most intriguing microcirculatory hemodynamic property. Though much is yet to be learned about this phenomena, the flow variability in individual capillaries is surely dependent on precapillary vasomotion, postcapillary dynamics, and structure, and most probably on the presence of arteriolar and venular arcuate interanastomosing

III. The Significance of Design, Cells, and Capillary Flow

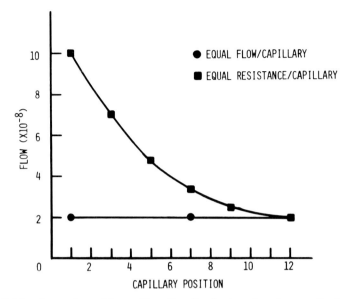

Fig. 8.21. Comparison of the capillary flow distribution when the most distal capillaries are constrained to have a certain minimum flow.

patterns. Flow intermittency has been observed to be related to the transient passage of white blood cells through the capillary network. It has also been proposed that flow intermittency is dependent on the assumed random passage of erythrocytes into one branch as opposed to its companion; i.e., dependent on the statistical distribution of the erythrocyte size and the distal flow field. The way in which each or all of these factors influences capillary flow depends on the vessel structure and dimensions and the topographical details of the vascular network in which they occur. Though this important aspect has not been studied in detail, it is illuminating to examine the differences in behavior of the capillary bed when two topographically extreme vascular distributions are considered, e.g., a dichotomous branching system and a side branching system (see Fig. 8.22). Analysis of events in these structures show that in the side branching pattern, the flow in a given capillary is usually much more sensitive to conditions in other capillaries. For example, if the flow in one of the capillaries is stopped (or reduced) because of an event or process within or distal to it, the flow in the remaining capillaries will increase. With reference to Fig. 8.22, if the flow in capillary 1 is stopped or reduced, the flow in capillary 4 will increase, whereas under the same circumstances, the flow of capillary 4 of the dichotomously branching system will remain unchanged. Because of the relatively noninteractive property of the dichotomously branching system, capillary flow is influenced by events principally in adjacent capillaries, whereas in the side branching configuration all

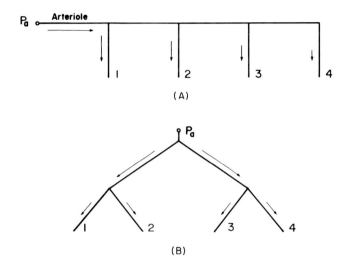

Fig. 8.22. Section of terminal microvasculature represented by side branching pattern (A) and dichotomously branching pattern (B). Flow in a given capillary in the side branching pattern is much more dependent on conditions in the other capillaries than is the case with dichotomous branching.

capillary flows are dependent on the conditions in all other capillaries having a common parent vessel. If one were to introduce interconnections between arterioles and interconnections between postcapillary venules, the interaction of both of these structures (dichotomous and side branching) would increase. Further insight into the relative roles of arteriole vasomotion, cellular influence, and vascular topology on the temporal and spacial variability of capillary blood flow must await further *in vivo* data, companion analyses, and interpretation of this data.

B. Cellular Components and Capillary Flow

If the fluid circulating through our blood vessels was devoid of erythrocytes and leukocytes, then analysis and understanding of microvascular hemodynamics would be much less difficult than it actually is. Pressure and flow relationships in individual vessels could be handled using classical steady flow equations (Poiseuille or Reynolds), and questions concerning inertia, convective acceleration, or significant pulsatility could be approached by using the appropriate versions of the Navier–Stokes equations (Navier, 1823; Stokes, 1845) suitably modified (Womersely). Provided that sufficient topographical information was available, quantitative descriptions of pressure–flow in the vascular network could be coupled to provide an overall description of its general hemodynamics,

III. The Significance of Design, Cells, and Capillary Flow

as described in the preceding sections. Problems associated with hemodynamic events at branch points would still present significant difficulties, as would a suitable approach to handling the complex vascular topology. But fortunately for us, though perhaps unfortunately for the hemodynamicist, erythrocytes, leukocytes, and other cellular components are present. The complexity of microvascular hemodynamics is largely due to this fact.

Putting aside these complexities for a moment, many aspects of the properties of blood flow in the microcirculation may be understood by first considering the forces which are acting on an infinitesimally small fluid volume or "fluid particle." Velocity of the fluid particle is governed by a balance between the net force acting on the particle that tends to produce motion and the inertial and viscous forces that tend to resist motion.

The inertial force arises because of the fluid mass and because changes in fluid velocity constitute changes in momentum, which must be supplied by the driving force. In general, there are two kinds of velocity changes that contribute to this inertial resisting force. One is due to the time-varying nature of flow associated with the heartbeat as well as to local microcirculatory mechanisms such as vasomotion. Because of this temporal variation, the velocity will experience a time rate of change. However, even in the absence of an intrinsic time variation there is another factor which may cause a rate of change in the velocity of an "observed" fluid particle. This is the geometry of the blood vessels themselves. If blood is flowing in an unbranched vessel of uniform diameter, then there would be no change in the velocity of a fluid particle moving through that vessel. However, since in most vessels diameter is not uniform through its length, velocity changes will be experienced as a consequence of flow continuity. Further, at each branch site there would be additional changes in velocity. The net effect of these combined inertial factors is to require the presence of a pressure gradient sufficient to counteract them so that flow may be maintained.

The viscous resistive force arises because of the interaction between adjacent fluid particles and the blood vessel wall. The net result of this interaction is to produce a resistive force that is proportional to velocity variation across the vessel diameter. Just as with the case of the inertial force in which there is a physical parameter (mass or inertia), there is a corresponding constant of proportionality which relates velocity to the viscous force. This constant is called the coefficient of viscosity or simply viscosity.

The most concise mathematical statement of these relationships is given by a vector form of the Navier-Stokes equation.

$$\nabla P = \underbrace{\rho \frac{D\vec{v}}{Dt}}_{A} + \underbrace{\mu \nabla^2 \vec{v}}_{B} \tag{3}$$

Since ρ is the fluid density (mass/volume), Term A represents inertial force per volume associated with an accelerating fluid "particle" of infinitesimally small volume. Physically, the acceleration is that which would be perceived if one were riding along with the particle, and it may be due to an intrinsic rate of change in particle velocity (v) with time and/or a rate of change of velocity due to dimensional changes in the fluid path. In Term B, the coefficient of viscosity is given by μ and the entire term expresses the viscous force per unit volume. The net effect of the presence of these forces is to have a situation in which the pressure–flow relationships are governed by a balance between the pressure gradient on the left hand and the inertial and viscous force on the right hand.

Equation 3 applies to blood flow in large arteries as well as in smaller ones, but it does not intrinsically deal with the problem of significant flow heterogeneity found in the microcirculation. The hemodynamic impact of the presence of the cellular components varies with the size of the vessel being studied and becomes increasingly significant as the vessel diameter of interest becomes smaller. One of the most challenging problems to both experimentalist and theoretician is the quantitative description of the role of these cellular components in capillaries and in those vessels that have internal diameters which are less than or close to the undistorted effective diameter of the red blood cell itself. A natural first question to pose in this connection is; "in what ways do the presence of red blood cells in the capillaries affect the pressure–flow relationships?" Current information bearing on this question has been gained principally from theoretical analyses and modeling experiments.

One approach to the modeling work is referred to as large-scale modeling. Using this method, an attempt is made to construct large model cells which mimic the physical properties of the red blood cells. These cells (rigid or deformable) are then caused to flow in large tubes under conditions in which the hydrodynamic parameters in the simulated capillary are dynamically similar to those in the true capillaries. The advantages of a large, scaled-up model of the capillary from the viewpoint of systematic testing and ease of obtaining hemodynamic measurements are obvious.

The theoretical basis of this method lies in the fact that the general hemodynamic equation (e.g., Eq. 3) can be used to describe the dynamic conditions of any size vessel, provided that appropriate boundary conditions and applicable assumptions are included. In practice, the dependent and independent variables are replaced by suitable dimensionless quantities (e.g., X/D), such that the experimental results obtained from a properly scaled model are directly relatable to true capillary flow. The principal assumption usually made is that the inertial component of flow resistance is negligible with respect to the viscous component. This is based on the fact that the capillary blood velocity is low and all characteristic dimensions (vessel diameter, RBC size) of interest are small. Under this condition, the flow field is governed by a balance between the gradient of pressure and the prevailing viscous shear forces.

III. The Significance of Design, Cells, and Capillary Flow

The approach most widely adopted to deal with the issue of RBC effects on the pressure–flow in capillaries is to consider the additional pressure drop caused by the presence of the cells. To this end, the total pressure drop ΔP along a capillary (model or actual) of length L and diameter D is expressed as the sum of two quantities; one associated with a flow of a Newtonian fluid simulating plasma through the capillary $\Delta P'$, and the other a quantity, ΔP^*, which represents the additional pressure drop associated with the red blood cells.

Using a variety of rigid simulated cell shapes, Sutera and Hochmuth (1968) demonstrated that for a given cell thickness (t) one of the important factors influencing the magnitude of ΔP^* was the ratio of the cell diameter (d) to tube diameter (D). For noninteracting spherical caps with d/D of 0.975, calculations reveal that each cell caused an additional pressure drop equal to about 4% of the simulated plasma pressure drop. For example, if 25 RBC's are present in a capillary and each satisfies the d/D and d/t model simulation, then the model results would predict the total capillary pressure drop to be twice of that expected from plasma flow alone.

The actual number of RBC's in a true capillary at any given time is dependent on a variety of factors and is quite variable. However, since there is evidence that the *in vivo* d/D ratio varies inversely with flow rate, the strong dependence of ΔP^* on d/D may give rise to a significant nonlinearity in the capillary pressure–flow relationship. The significance of this important component, red cell flexibility, was dealt with by Lee and Fung (1969) using thin-wall rubber models of RBC's in a large scale fluid dynamic simulation. Using cells that, in some instances, were larger than the simulated capillaries, the nonlinear relationship between capillary pressure gradient and flow was experimentally confirmed and clearly associated with the presence of the simulated red cells.

Since the exact properties of the erythrocyte membrane are not known, the simulation of the RBC is perhaps the weakest link in the large scale modeling procedures. Sutera *et al.* (1970) employed model cells similar in design to those used by Lee and Fung but having substantially thinner and more flexible membranes. Using high speed photography, they showed that the deformation of these model cells is similar to that seen when real blood cells are caused to flow in 4–10 μm glass capillary tubing (Hochmuth *et al.*, 1970; Seshadri *et al.*, 1970). The results of these studies suggested that because of the cell flexibility, the actual RBC is deformed by an amount which depends on the cell velocity. The effect of this deformation is to cause the d/D ratio to decrease with increasing flow over a range of about 0.2–1.0 mm/sec. Measured additional pressure drops of the model cells decreased nonlinearly with increasing deformation as the flow parameter was increased. Thus, the concept of a nonlinear capillary resistance to blood flow due to the presence of RBC's became reasonably well accepted.

The nonlinear resistance associated with capillary flow had to a certain extent been predicted by Lighthill (1968, 1969) and elaborated on by Fitz-Gerald

(1969a,b) by applying the analytical techniques of lubrication theory. Originally this theory was developed to analyze the hydrodynamics within and surrounding an oil film (or other lubricating agents), which permitted one surface to slide past another while a large force was transmitted between them. Lighthill argued that a similar process must be at work when flexible RBC's are deformed in capillary flow and equated the action of the thin plasma layer between the cell and the wall with the function of a lubricating agent. Accordingly, the required radial stress to deform the cell would be transmitted from the wall to the cell through the thin plasma layer by viscous action. It is this viscous component in the film that gives rise to a pressure variation along the length of the lubrication layer and hence along the surface of the deformed cell. Associated with this thin plasma layer is a viscous force tending to offer resistance to the RBC motion by an amount which is directly proportional to the cell velocity and inversely proportional to the plasma layer thickness. However, since lubrication theory indicates that for the flexible red cell, the plasma layer itself varies as the square root of the cell velocity, the conclusion reached is that the additional pressure drop will be proportional to the square root of the cell velocity. The result implies that capillary flow resistance is nonlinear and increases with decreasing red cell velocity.

A possible implication of such a nonlinear resistance property is that when a terminal arteriole bifurcates into two capillaries, the capillary with the larger initial velocity will tend to receive an increasingly larger fraction of the erythrocyte flow. Such behavior has been demonstrated *in vitro* using models (Fung, 1973), but a systematic test of this concept *in vivo* remains to be done. Another consequence of this predicted nonlinear resistance property is the possibility that if the velocity in a capillary continues to decrease, there may be a point at which flow ceases entirely for a nonzero pressure gradient. Addressing this issue in part, Lingard (1974) simultaneously measured the pressure gradient and red cell velocity of human erythrocytes caused to flow through capillary pores which were 7 μm in diameter and 13 μm in length. Analysis of these data shows that for this experimental arrangement, the onset of the nonlinear increase in resistance occurs at about a cell velocity of 1 mm/sec. For cell velocities less than this value, capillary flow resistance was found to increase nonlinearly, but then abruptly to level off and remain constant for cell velocities less than about 0.3 mm/sec. These results were later confirmed for 5 μm pores (Lingard, 1977a,b), and the phenomena was attributed to a velocity-dependent change in the effective width of the thin film between the red blood cell and tube wall associated with RBC deformation. The basic nonlinear property and its dependence on mean cell velocity was also found when using capillary arrays with a length of 500 μm (Lingard, 1979), although the magnitude of the low velocity relative resistance was less than that with shorter pores. The issues of the role of lubrication phenomena, nonlinear resistance associated with capillary flow, and the concept of "Sieze up" are controversial at this writing. In part, this is due to the fact that

IV. Other Factors Influencing *In Vivo* Hemodynamics

these properties are difficult to study *in vivo*, and thus all information is based on theoretical and *in vitro* studies. Finally, if present *in vivo*, it is not known whether the magnitude of the nonlinear characteristic is sufficient to cause any significant hemodynamic effects.

IV. OTHER FACTORS INFLUENCING *IN VIVO* HEMODYNAMICS

In its journey to the capillaries, the blood must first negotiate a tortuous and often complex highway system from the large distributing arteries to the terminal arterioles. Within the vascular bed the arteriolar system is seen to be comprised of a heterogenous branching system characterized by a progressive diminution in vessel diameters, when possible conceptualized as a set of anatomically or functionally distinctive branching orders. The overall microcirculatory hemodynamics are dependent on the geometry and dimensions of each vessel, the spacial and topographical relationship of vessel to vessel, and branching order to branching order.

Within each vessel, the pressure–flow relationships are dependent on the physical characteristics of the red blood cells and their concentration and distribution within the vessel lumen. Classically, the blood component of the pertinent hemodynamic equations is accounted for by specifying the viscosity, η. Thus, the simplest characterization of the resistance to blood flow in a single vessel includes two components; a geometric factor (L/D^4) and a fluid factor (η). The well known Poiseuille formula incorporates both of these factors and states that the resistance to blood flow R of a cylindrical vessel with length L and diameter D is given by:

$$R = \frac{128}{\pi} \frac{L}{D^4} \eta \qquad (4)$$

The viscosity of the blood (or its "apparent" viscosity) is thus seen to be a measure of the resistance to flow associated with the blood itself (which turns out to be dependent on the vessels in which it flows). The study of the deformation and flow properties of fluids falls under the general heading of rheology, and those studies directed toward the understanding of the same in blood are termed biorheology or hemorheology. Many excellent detailed reviews are available (Copley, 1974; Baez, 1977; Braasch, 1971; Cokelet, 1978). However, there are certain elements falling into this general category which so directly influence *in vivo* microcirculatory hemodynamics that a brief discussion of them is warranted.

One which may be of particular importance both with regard to hemodynamics as well as to oxygen delivery (Klitzman and Duling, 1979) is the fact that the volume concentration of RBC's (hematocrit) is a function of both the vessel size and the position of the vessel within the microvascular tree. Substantial *in vivo* data indicates that there is a progressive diminution in hematocrit through the vascular bed, and capillary hematocrits as low as 8% have been reported (Lipowsky and Zweifach, 1977). One of the most widely held theories that could account for this hematocrit reduction is based on the concept of plasma skimming. This term describes the process in which the RBC concentration in a branch is less than that in its parent vessel, in part because of a nonuniform RBC radial distribution in the parent vessel. Because of the combination of mechanical and hydrodynamic exclusionary effects, the region closest to the vessel wall has less RBC's per unit volume present than does the remainder of the vessel. Since the blood which is drawn into a branch preferentially is drawn from this "peripheral plasma layer," the branch will have a lesser RBC concentration than its parent. Qualitatively similar processes occurring at each branching order could thus account for the vascular bed hematocrit profile.

Though conceptually satisfying, the plasma skimming theory is probably only part of the story. Theoretical analyses of blood flow in tubes having microvessel dimensions show that the mean volume concentration one would measure in a vessel depends on the flow conditions within the vessel itself (Thomas, 1962). Since these flow conditions depend on a large number of interactive factors (vessel diameter, magnitude of blood velocity, velocity radial profile, and RBC concentration profile), a unified theory has not yet evolved. However, significant inroads into this complex issue have been made. By taking into account expected differences in the flow regime and the size of the microvessels, Whitmore (1967) put forward a theory which predicted the presence of a hematocrit minimum to occur in vessels between 7 and 16 μm. Barbee and Cokelet (1971a), using well controlled *in vitro* conditions, provided quantitative data to describe the hematocrit reduction in tubes from 221 to 29 μm diameter, and they showed (Barbee and Cokelet, 1971b) that pressure–flow relationships could be predicted if the actual vessel hematocrit were known. Comparisons of theoretical and *in vitro* studies (Lin *et al.*, 1973), as well as more recent studies (Yen and Fung, 1977), suggest that a diameter dependent minimum in viscosity as previously put forward may be present.

The overall characterization of the *in vivo* implications of these and a variety of other complex interactive phenomena (e.g., local effects at branch points, pulsatility) are just beginning. The simultaneous measurement of blood pressure gradients and blood flow *in vivo,* recently applied to the cat mesentery (Lipowsky *et al.*, 1978), may foreshadow a new era in the study of microcirculatory hemodynamics. Using these techniques together with the powerful analytical methods that are now available, it should be possible to obtain the detailed experimental information to settle an array of long standing controversial issues and to determine which aspects of microcirculatory hemodynamics are simply academic curiosities and which are physiologically and functionally significant.

REFERENCES

Baez, S. (1977). Microcirculation. *Annu. Rev. Physiol.* **39**, 391–415.
Barbee, J. H., and Cokelet G. R. (1971a). The Fahraeus effect. *Microvasc. Res.* **3**, 6–16.
Barbee, J. H., and Cokelet, G. R. (1971b). Prediction of blood flow in tubes with diameters as small as 29μ. *Microvasc. Res.* **3**, 17–21.
Bohlen, H. G., and Gore, R. W. (1977). Comparison of microvascular pressures and diameters in the innervated and denervated rat intestine. *Microvasc. Res.* **14**, 251–264.
Bohlen, G. H., and Gore, R. W. (1979). Microvascular pressures in rat intestinal muscle during direct nerve stimulation. *Microvasc. Res.* **17**, 27–37.
Bohlen, G. H., Gore, R. W., and Hutchins, P. W. (1977). Comparison of microvascular pressures in normal and spontaneously hypertensive rats. *Microvasc. Res.* **13**, 125–130.
Braasch, D. (1971). Red cell deformability and capillary blood flow. *Physiol. Res.* **51**, 679–696.
Cohn, D. I. (1954). Optimal systems, I. The vascular system. *Bull. Math. Biophys.* **16**, 59–74.
Cohn, D. L. (1955). Optimal systems II: The vascular system. *Bull. Math. Biophys.* **17**, 219–227.
Cokelet, G. (1978). *In* "Peripheral Circulation" (Paul C. Johnson, ed.), Wiley, New York. pp. 81–110.
Copley, A. L. (1974). Hemorheological aspects of the endothelium-plasma interface. *Microvasc. Res.* **8**, 192–212.
Eriksson, E., and Myrhage, R. (1972). Microvascular dimensions and blood flow in skeletal muscle. *Acta Physiol. Scand.* **86**, 211–222.
Fitz-Gerald, J. M. (1969a). Mechanics of red cell motion through very narrow capillaries. *Proc. R. Soc. London, Ser. B* **174**, 193–227.
Fitz-Gerald, J. M. (1969b). Implications of a theory of erythrocyte motion in narrow capillaries. *J. Appl. Physiol.* **26**, 912–918.
Fronek, K., and Zweifach, B. (1974). Pre- and post-capillary resistances in cat mesentery. *Microvasc. Res.* **7**, 351–361.
Fronek, K., and Zweifach, B. W. (1975). Microvascular pressure distribution in skeletal muscle and the effect of vasodilation. *Am. J. Physiol.* **228**, 791–796.
Fronek, K., and Zweifach, B. (1977). Microvascular blood flow in cat tenuissimus muscle. *Microvasc. Res.* **14**, 181–189.
Fung, Y. C. (1973). Stoichastic flow in capillary blood vessels. *Microvasc. Res.* **5**, 35–48.
Gore, R. W. (1974). Pressures in cat mesenteric arterioles and capillaries during changes in systemic arterial blood pressure. *Circ. Res.* **34**, 581–591.
Gross, J. F., and Intaglietta, M. (1973). Effects of morphology and structural properties on microvascular hemodynamics. *Bibl. Anat.* **11**, 532–539.
Gross, J. F., Intaglietta, M., and Zweifach, B. (1974). Network model of pulsatile hemodynamics in the microcirculation of the rabbit omentum. *Am. J. Physiol.* **22**, 1117–1123.
Hochmuth, R. M., Marple, R. N., and Sutera, S. P. (1970). Capillary blood flow, 1. Erythrocyte deformation in glass capillaries. *Microvasc. Res.* **2**, 409–419.
Intaglietta, M., and Zweifach, B. (1971). Geometrical model of the microvasculature of rabbit omentum from *in vivo* measurements. *Circ. Res.* **28**, 593–600.
Intaglietta, M., Richardson, D. R., and Tompkins, W. R. (1971). Blood pressure, flow and elastic properties in microvessels of cat omentum. *Am. J. Physiol.* **221**, 922–928.
Jeffords, J. V., and Knisely, M. (1956). Concerning the geometric shapes of arteries and arterioles. *Angiology* **7**, 105–136.
Klitzman, B., and Duling, B. R. (1979). Microvascular hematocrit and red cell flow in resting and contracting striated muscle. *Am. J. Physiol.* **237**, H481–H490.
Krogh, A. (1929). "The Anatomy and Physiology of Capillaries." Yale Univ. Press, New Haven, Connecticut. (Reprint, Hafner, New York, 1959.)

Lee, J. S., and Fung, Y. C. (1969). Modeling experiments of a single red blood cell moving in a capillary blood vessel. *Microvasc. Res.* **1,** 221–243.

Lighthill, J. J. (1968). Pressure forcing of tight-fitting pellets along fluid-filled elastic tubes. *J. Fluid Mech.* **34,** 113–143.

Lighthill, M. D. (1969). Motion in narrow capillaries from the standpoint of lubrication theory. *In* "Circulating and Respiratory Mass Transport" (G. E. W. Wolstenholme and J. Knight, eds.), pp. 85–96. Churchill, London.

Lin, K. L., Lopez, L., and Hellums, J. D. (1973). Blood flow in capillaries. *Microvasc. Res.* **5,** 7–19.

Lingard, P. S. (1974). Capillary pore rheology of erythrocytes I. Hydroelastic behavior of human erythrocytes. *Microvasc. Res.* **8,** 55–63.

Lingard, P. S. (1977a). Capillary pore rheology of erythrocytes. III. On the interpretation of human behavior in narrow capillary pores. *Microvasc. Res.* **13,** 29–58.

Lingard, P. S. (1977b). Capillary pore rheology of erythrocytes IV. Effect of pore diameter and hematocrit. *Microvasc. Res.* **13,** 59–77.

Lingard, P. S. (1979). Capillary pore rheology of erythrocytes V. The glass capillary array—effect of velocity and hematocrit in long bore tubes. *Microvasc. Res.* **17,** 272–289.

Lipowsky, H. H., and Zweifach, B. W. (1974). Network analysis of microcirculation of cat mesentery. *Microvasc. Res.* **7,** 73–83.

Lipowsky, H. H., and Zweifach, B. W. (1977). Methods for the simultaneous measurement of pressure differentials for rheological studies. *Microvasc. Res.* **14,** 345–361.

Lipowsky, H. H., Kovalcheck, S., and Zweifach, B. W. (1978). The distribution of blood rheological parameters in the microvasculature of cat mesentery. *Circ. Res.* **43,** 738–749.

Ma, Y. P., Koo, A., Kwan, H. C., and Cheng, K. K. (1974). On-line measurement of the dynamic velocity of erythrocytes in the cerebral microvessesl in the rat. *Microvasc. Res.* **8,** 1–13.

Mayrovitz, H. N. (1974). The Microcirculation: Theory and Experiment. Ph.D. Thesis, University of Pennsylvania, Philadelphia, Pennsylvania.

Mayrovitz, H. N., and Wiedeman, M. P., and Noordergraaf, A. (1975). Microvascular hemodynamic variations accompanying microvessel dimensional changes. *Microvasc. Res.* **10,** 322–339.

Mayrovitz, H. N., Wiedeman, M. P., and Noordergraaf, A. (1976a). Analytical characterization of microvascular resistance distribution. *Bull. Math. Biol.* **38,** 71–82.

Mayrovitz, H. N., Tuma, R. F., and Wiedeman, M. P. (1976b). Pulsatility of microvascular blood velocity. *Fed. Proc. Fed. Am. Soc. Exp. Biol.* **35,** 233.

Mayrovitz, H. N., Tuma, R. F., and Wiedeman, M. P. (1977). Relationship between microvascular blood velocity and pressure distribution. *Am. J. Physiol.* **232,** H400–H405.

Mayrovitz, H. N., Wiedeman, M. P., and Noordergraaf, A. (1978). Interaction in the microcirculation. *In* "Cardiovascular System Dynamics" (J. Baan, A. Noordergraaf, and J. Raines, eds.). MIT Press, Cambridge, Massachusetts.

Navier, C. L. (1823). *Mem. Acad. R. Sci. Lett. Belg.* **6,** 389–416.

Nicoll, P. A. (1969). Intrinsic regulation in the microcirculation based on direct pressure measurements. *In* "Microcirculation (W. L. Winters and A. N. Brest, eds.), pp. 89–101. Thomas, Springfield, Illinois.

Roy, C. S., and Brown, J. G. (1879). The blood pressure and its variations in arterioles, capillaries and smaller veins. *J. Physiol. (London)* **2,** 323–359.

Seshadri, V., Hochmuth, R. M., Croce, P. A., and Sutera, S. P. (1970). Capillary blood flow III, Deformable model cells compared to erythrocytes *in vitro*. *Microvasc. Res.* **2,** 434–442.

Skalak, R., and Statitis, T. (1966). A porus tapered elastic tube model of a vascular bed. *Biomechanics* **28,** 57–65.

Stokes, C. G. (1845). *Trans. Cambridge Philos. Soc.* **8,** 287–305.

References

Sutera, S. P., and Hochmuth, R. M. (1968). Large scale modeling of blood flow in the capillaries. *Biorheology* **5**, 45–73.

Sutera, S. P., Seshardri, V., Croce, P. A., and Hochmuth, R. M. (1970). Capillary blood flow II, Deformable model cells in tube flow. *Microvasc. Res.* **2**, 420–433.

Thomas, H. W. (1962). The wall effect in capillary instruments. *Biorheology* **1**, 41–56.

Walawender, W. P., and Chen, T. Y. (1975). Blood flow in tapered tubes. *Microvasc. Res.* **9**, 190–205.

Wayland, H., and Johnson, P. C. (1967). Erythrocyte velocity measurement in microvessels by a two-split photometric method. *J. Appl. Physiol.* **22**, 333–337.

Whitmore, R. L. (1967). A theory of blood flow in small vessels. *J. Appl. Physiol.* **22**, 767–771.

Wiederhielm, C. A., and Weston, B. V. (1973). Microvascular lymphatic and tissue pressures in the unanesthetized mammal. *Am. J. Physiol.* **225**, 992–996.

Yen, R. T., and Fung, Y. C. (1977). Inversion of Fahraeus effect and effect of mainstream flow on capillary hematocrit. *J. Appl. Physiol.*, **42**, 578–586.

Zamir, J. (1977). Shear forces and blood vessel radii in the cardiovascular system. *J. Gen. Physiol.* **69**, 449–461.

Zweifach, B. W. (1974a). Quantitative studies of microcirculatory structure and function I, Analysis of pressure distribution in the terminal vascular bed in cat mesentery. *Circ. Res.* **34**, 843–857.

Zweifach, B. W. (1974b). Quantitative studies of microcirculatory structure and function II, Direct measurement of capillary pressure in splanchnic mesenteric vessels. *Circ. Res.* **34**, 856–866.

Zweifach, B. W., and Lipowsky, H. H. (1977). Quantitative studies of microcirculatory structure and function III, Microvascular hemodynamics of cat mesentery and rabbit omentum. *Circ. Res.* **41**, 380–390.

Index

A

Absorption, intestinal villi, 127–129
Adenine nucleotide, vasodilatory effects, 117
Adenosine
 cardiac blood flow effects, 122, 136
 cerebral blood flow effects, 121
 vasodilatory effects, 105, 117
Adenosine triphosphate, 130
Adrenergic nerve
 blood flow control, 100–101, 112–116
 skeletal muscle innervation, 112–113
Adrenergic receptor
 blockage, 122
 cardiac, 122
 vascular smooth muscle, 111–112
Adventitia, pulmonary, 51
Aggregation, platelet, 10, 66
Alimentary canal, microcirculation, 123–129
Alveoli, 52
Amine, *see also* specific amines
 vasoactivity, 108
Anatomical shunt, *see* Preferential channel
Anastomosis
 arterio-arterial
  cardiac, 123
 arteriovenous
  absence in brain, 56
  blood flow, 66
  contraction, 64–65

 cutaneous, 129
 diameter, 63
 gastric, 35
 skeletal muscle, 28–29, 119
 cremaster muscle, 24–25
 intestinal, 37
Anatomy and Physiology of Capillaries, 10
Anesthesia, effect on vasomotion, 56, 75–76
Anesthetization, of laboratory animals, 75–76, 77
Angiotensin, 110, 117
Arachadonic acid, 105–106
Arc-capillary, 28, 29
Arterial-venous bridge, 13–14
Arterial vessel, *see also* Arteriole; Artery
 branching order, 63
 cerebral, 55, 56
 contraction, 63–64
 diameter, 55, 56
 innervation, 51, 116
 mesenteric, 39, 42
 rabbit ear, 63–64
 splenic, 51
Arteriole, 17, 18
 bat wing, 67
 blood flow, 45, 53, 55, 65
  input impedance, 197, 198, 199, 200, 201
  rabbit ear, 63
 blood pressure, 27
 branching order
  bulbar conjunctiva, 55

217

Arteriole, branching order (*continued*)
   cardiac, 30, 31
   gastric, 33
   hemodynamic function, 195–206
   intestinal, 35, 37
  bulbar conjunctiva, 54, 55
   cardiac, 30
   cerebral, 55
    innervation, 120
   contraction, 53, 54
   cremaster muscle, 23
   diameter, 39, 50, 53, 54, 59, 67, 107
   hamster cheek pouch, 59, 61
   hemodynamics, 195–206
   hepatic, 47–48
   innervation, 61, 100, 120
   intestinal, 35, 37
   intraluminal pressure effects, 107
   mesenteric, 42
   neural control, 113
   pulmonary, 52, 53
   rabbit ear, 63
   terminal, 17, 18
    bat wing, 67
    bulbar conjunctiva, 54
    definition, 17
    diameter, 190
    dilation, 102
    hamster cheek pouch, 59
    hepatic, 48
    innervation, 100, 101–102
    mesenteric, 42
    pulmonary, 52
Arteriole-capillary distribution, 195–206
Arteriovenous anastomosis, *see* Anastomosis, arteriovenous
Artery
  bat wing, 67
  blood flow, 35, 45, 63
  blood pressure, 27
  branching, 33, 34, 59, 63, 67
  carotid, 57, 61
  cerebral, 55–56
  cremaster, 22–23, 24–25
  diameter, 67
  dilation, 25
  gastric, 33
  gastroepiploic, 33
  inferior labial, 58
  innervation, 100
  mesenteric, 35, 39, 42, 45
  mucosal, 33, 34–35, 37
  perforating, 56
  pulmonary, 51
  small, 67, 100
  spermatic, 22–23
  superior mesenteric, 35, 39
  superior saccular, 57–58
  tenuissimus muscle, 25
Artery to artery connection, 56
ATP, *see* Adenosine triphosphate
Autonomic nervous system, 117, 120, 122
Autoregulation, blood flow, 131–137
Autoregulatory escape, blood flow, 124, 126, 136–137
A-v bridge, *see* Arterial-venous bridge
Axonal reflex, 130

B

Barbiturate, 76, 77
Bat, *see also Myotis lucifugus*
Bat, little brown, *see Myotis lucifugus*
Bat wing
  anatomy, 66–67
  blood velocity, 188, 189
  microvascular observation, 92–94
  α-receptor, 117
  vasculature, 66–70, 191–192
Biorheology, 211
Blood
  molecular transport, 140–152
  oxygen exchange, 143
  solute diffusion, 142–145
Blood-borne factor, *see* Humoral factor
Blood flow, *see also* Microcirculation
  arterial, 35, 45, 47–48, 57–59, 63
  arteriole, 45, 55, 63
  autoregulation, 131–137
  bulbar conjunctiva, 55
  capacity, 67
  capillary, 25, 28, 29–30, 45, 68, 102, 108, 115–116, 117–119, 191, 194–206, 211
   cellular components effects, 206–211
   hemodynamics, 206–211
   sympathetic nerve stimulation, 113–114
   variability, 204–205
  cardiac, 90–91, 136
   control, 122–123

Index

cerebral, 55–56, 120, 121, 132–133, 134
   carbon dioxide effects, 120, 121
   control, 119–122, 136
   reversal, 55, 56
early research, 3–11
gastric arteries, 35
gastrointestinal mobility effects, 126–127
gastrointestinal tract
   control, 123–129, 134–135
hamster cheek pouch, 57–59
humoral control, 108–112, 126–127
input impedance, 196–197, 198, 199–200, 201
leukocyte obstruction, 66
metabolic control, 103–106, 126–127
microvascular regulation, 99–112
mucosal, 78–79
myogenic control, 106–108, 127, 136
neural control, 100–103, 112–116, 119–120, 124, 126, 136
oxygen effects, 103–105
phasic, 59
precapillary resistance, 27, 103–104
pulmonary, 53
rabbit ear, 63–65
regulation, 68, 99–137
skeletal muscle
   autoregulation, 131–132
   control, 112–119
skin, 129–131
splenic, 48–51
spinotrapezius muscle, 119
sympathetic nervous system control, 112–116
temperature effects, 131
tenuissimus muscle, 119
velocity, 27, 76
   arteriole, 54
   capillary, 188
   equation, 163, 166
   measurement, 157–169, 186–190
   profile, 188
venule function, 37–38
volume, 165–166
Blood plasma
   skimming theory, 212
   velocity, 165, 166
Blood pressure
   anesthesia effects, 76
   arterial, 27
   arteriole, 27

capillary, 190, 202–204
cerebral, 55
diastolic, 170
hamster cheek pouch, 59–61
intestinal, 39
measurement, 9–10, 169–174, 182–186, 188–190
microvascular
   distribution, 178–179, 180, 181, 186
   measurement, 182–186, 188–190
   systemic pressure correlation, 184–185
precapillary, 27
profile, 184–185, 186
renin effects, 110
systemic, 59–61, 136
   microvascular pressure correlations, 184–185
   variations, 184–185
   vascular resistance effects, 114
systolic, 170
vessel diameter effects, 39, 42–43, 190
Blood vessel, see also specific types of blood vessel
   anesthesia effects, 75–76
   catecholamine sensitivity, 112
   identification by blood flow, 63
   microscopic observation, 63–70
   pressure profile, 184
   resting tone, 59
   taper, 199, 200, 201
Blood viscosity, 189, 211
Bradykinin, 110, 130
Brain, see also Pia mater
   blood flow, 55–57
      autoregulation, 132–133, 134
      control, 119–122, 136
   blood pressure, 55, 120
   innervation, 119–120
   vascular pattern, 55
Branching order
   arterial, 33, 34, 59, 63, 67
   arteriole, 30, 31, 33, 35, 37, 55, 195–206
   bat wing, 67
   blood velocity, 187–188
   bulbar conjunctiva, 55
   microvasculature, 178–179, 182, 183, 184
   model, 192–193
   pressure profile, 185
   rabbit ear, 60–61, 63

Bronchiole, 51
Bulbar conjunctiva, 53–55

C

Capillary
  blood flow, 25, 28, 29–30, 66, 68, 102, 108, 115–116, 117–119, 191, 194–206
    effect of cellular components, 206–211
    hemodynamics, 206–212
    input impedance, 197
    neural stimulation, 113–114
    variability, 204–205
    velocity, 188
  blood pressure, 190, 202–204
  definition, 17
  density, 194–195
  diameter, 39, 63, 67, 202
  diffusion pore, 143–144, 150–151, 152
  discontinuous, 140–141
  endothelial vesicle, 147–152
  fenestrated, 140, 150
  flow resistance, 208–211
  fluid exchange, 76
  fluid filtration, 108, 145–147
  hydrostatic pressure, 127
  nonfenestrated, 140
  perfusion, 117–119, 123, 126
  permeability, 76, 110, 143–144
  pulmonary, 52
  tenuissimus muscle, 28
  vascular resistance, 196, 197, 202–205
Capillary-arteriole distribution, *see* Arteriole-capillary distribution
Capillary bed
  bat wing, 68
  bulbar conjunctiva, 55
  cerebral, 56
  cremaster muscle, 23–24
  defined, 42
  gastric, 33–34
  gastrointestinal tract, 37, 127–128
  mesenteric, 42
  spleen, 50, 51
  structure, 12–20
Capillary filtration coefficient, 113, 114
Carbon dioxide, vasodilatory effects, 103, 117, 120–121, 136
Cardiac output, 76, 99
Cardiovascular system, anesthetic effects, 75–76
Cat, use in microvascular studies, 78, 86, 88–89
Catecholamine
  release, 76
  vasoactivity, 108, 111–112
Central nervous system, 100–103
Chloralose, 76, 79, 86
Cholecystokinin, 126
Cholinergic nerve
  arterial innervation, 116
  cutaneous blood flow control, 130
Contractility, *see also* Vasoconstriction
  of lymphatic vessels, 43, 69
Countercurrent mechanism, intestinal villi, 127–128
Cranial window technique, 88–89
Craniotomy, 55, 88
Cremaster muscle
  artery, 22–23
  blood pressure, 184, 185
  capillary bed, 23–24
  exteriorization, 81–82
  innervation, 25
  microvasculature, 21–25
    microscopic observation, 77, 81–82
  surgical preparation, 81–82

D

De moto cordis, 3
Denervation
  effect on arterial dilation, 25
  surgical technique, 79
Diabetes, 53, 55
Diapedesis, 66
Diastole, 134
Diffusion
  equation, 142–143
  lipid-insolubles, 143, 145
  lipid-solubles, 143, 144–145
Dilation, *see* Vasodilation

E

Electro-optical technique, of blood velocity measurement, 160–169
Endocardium, blood flow, 134

Index

Endothelial cell, 48
   membrane, 144-145
Endothelial vesicle, capillary, 147-152
Endothelium
   capillary, 50
   lymphatic, 63
Epicardium, blood flow, 134
Epinephrine
   adrenergic receptor stimulation, 111, 112
   vasoactivity, 117, 127
Erythrocyte
   and blood flow intermittency, 205
   capillary flow effects, 206, 211
   concentration, 212
   shape during flow, 66, 208-209, 210
   velocity measurement, 158-169, 187, 188
Ether, 76
Extensor hallucis proprius, 29
Exteriorization
   cremaster muscle, 81-82
   hamster cheek pouch, 80
   methodology, 77-82

F

Fenestrae, capillary, 140, 150
Filtration, capillary, 108, 145-147
   coefficient, 113, 114
Fluid component, effect on blood flow, 206-211
Fluid exchange
   microcirculatory, 140-152
   vesicular transport, 147-152
Fluid velocity, 207
Fluid viscosity, 207-208

G

Galen, 3
Gastrin, vasoactivity, 126
Gastrointestinal motility, effect on blood flow, 126-127
Gastrointestinal tract
   blood flow
      autoregulation, 134-135
      control, 123-129
   innervation, 126
Genitofemoral nerve, 25
Guinea pig, use in microvascular studies, 89

H

Hales, Stephen, 9-10
Hall, Marshall, 10
Hamster cheek pouch
   chamber, 83-84
   microscopic observation, 80, 83-84
   structure, 57
   surgical preparation, 80
   vasculature, 57-61
Harvey, William, 3-5
Heart
   blood flow
      autoregulation, 134
      control, 122-123
   innervation, 122
   microvasculature, 30-32, 122-123, 134
Heat
   effect on anastomosis contractility, 64-65
   effect on blood flow, 63
   effect on lymphatic vessels, 63
Hematocrit, 167-168, 212
Hemodynamics
   arteriole-capillary distribution, 195-206
   capillary blood flow, 206-212
   microcirculatory, 177-212
Hemorrhage, 184
Histamine, vasodilatory effects, 108, 109, 117, 130
Hormone, vasoactive, 126, 136
Humoral control
   blood flow, 108-112, 117, 126-127
      gastrointestinal, 126-127
      skeletal muscle, 117
Humoral factor, vasoactive properties, 108-112, 136
Hydrogen ion concentration, 117, 121
Hydrostatic pressure
   capillary, 127
   effect on capillary filtration, 145, 146, 147
5-Hydroxytrytamine, *see* Serotonin
Hypercapnia, 121
Hyperemia, 117, 122, 131
Hypertension, 133, 134, 184
Hypoxia, 121, 122, 126

I

Iliohypogastric nerve, 25
Ilioinguinal nerve, 25

Infection, effect on lymphatic vessels, 63
Innervation, vascular
 arterial, 51
 arteriole, 61, 100, 101-102, 120
 bat wing, 69
 cardiac, 122
 cerebral, 119-120
 colon, 126
 cremaster muscle, 25
 cutaneous, 129-130
 gastric, 126
 gastrointestinal, 124, 126
 hamster cheek pouch, 61
 pulmonary adventitia, 51
 skeletal muscle, 112-116
 sweat gland, 130
In situ organ preparation, 85-91
In situ tissue preparation, 77, 85-91
Intestine
 innervation, 124
 microscopic observation, 79-80
 vasculature, 35-39
Intraluminal pressure, 106-107
Isoproterenol, 79

J

Jones, T. W., 10

K

Kallikrein, 130
Kinin, 110
Krogh, August, 10

L

Laser-Doppler velocimetry, 168-169
Leeuwenhoek, Antoni van, 9, 10, 158
Leukocyte
 adherence, 65, 66, 76
 flow behavior, 65-66, 205
 migration, 63
 number, 76
 rigidity, 66
Little brown bat, *see Myotis lucifugus*

Liver
 microvasculature, 47-48, 90
 microscopic observation, 90
Lubrication theory, of capillary flow, 209-211
Lumen, occlusion, 48
Lung
 microvasculature, 51-53, 89-90
 microscopic observation, 89-90
Lymphatic vessel
 contractility, 69
 fluid filtration function, 147
 mesenteric, 43, 44
 rabbit ear, 63

M

Magnesium, vasoactivity, 117
Malpighi, Marcello, 4-8
Marshall, John, 9
Mesentery
 blood pressure, 184, 185
 cecal, 45-47, 77-78
 exteriorization, 77-78
 microvasculature, 39-47, 77-78
 velocity profile, 188
 venous system, 39, 42
Metabolic control
 blood flow
  autoregulatory, 131-133
  cardiac, 122
  cerebral, 120-121
  gastrointestinal, 126-127
  skeletal muscle, 117
  smooth muscle, 103-106
Metabolite, vasodilatory effects, 103, 117, 131
Metarteriole, 14, 15, 16, 17-18
 cremaster muscle, 25
Microcannula, 170, 171, 172, 173
Microcinematography, 90, 91, *see also* Photography
Microcirculation, *see also* Blood flow
 alimentary canal
 bulbar conjunctiva, 55
 cardiac, 122-123, 136
 cerebral, 55-56, 119-122, 132-133, 134, 136
 cutaneous, 129-131
 early research, 3-11
 fluid exchange, 140-152
 gastrointestinal, 123-129, 134-135

Index

hemodynamics, 177–212
humoral control, 108–112
metabolic control, 103–106, 126–127
myogenic control, 108–112
neural control, 100–103, 112–116
observation techniques, 75–94
pressure measurement, 169–174, 182–186, 188–190
pulmonary, 53
pulsatility, 188–190
regulation, 99–112, 135–136
skeletal muscle, 112–119
smooth muscle, 100–112
solute exchange, 140–152
velocity, 157–169, 186–190
Microscope, development, 8–9
Microscopic observation
microcirculation
bat wing, 66–70, 92–94
cardiac, 90–91
hepatic, 90
human skin, 65–66
methodology, 65–70, 75–94
pia mater, 88–89
rabbit ear, 63–65
spleen, 87–88
velocity, 157–169
Microthrombus, 66
Microvasculature
bat wing, 66–70
blood flow, *see* Microcirculation
branching, 178–179
bulbar conjunctiva, 53–55
cardiac, 122–123, 136
cremaster muscle, 21–25
extensor hallucis proprius, 29
gastric, 33–35
hemodynamics, 191–206
hepatic, 47–48
intestinal, 35–39
mesenteric, 39–43, 45–47
muscle, 21–32
pia mater, 55–57
pressure-flow distribution, 178–181
pulmonary, 51–53
rabbit ear, 63–65
sartorius muscle, 29–30
spinotrapezius muscle, 29
spleen, 48–51
tenuissimus muscle, 25–29
terminal, 194

vascular resistance, 178–181
visceral, 33–53
Migration, leukocyte, 63
Module, vascular, 42
Morphine, 76
Mucosa
arterial blood supply, 33, 34–35
vascular pressure, 127
Muller, Johannes, 10
Muscle, *see also* specific muscles
contraction, 118–119
microvasculature, 21–32
tone, 100, 102–103
Myocardium, 30–32
Myogenic control
arteriole response, 68–69
blood flow, 106–108, 131–133, 136
cardiac, 123
gastrointestinal, 127
skeletal muscle, 116–117
contractile activity, 101–102
Myotis lucifugus, use in microvasculature studies, 66–70

N

Navier–Stokes equation, 206, 207–208
Nerve, *see also* specific nerves
muscle tonic effect, 100, 102–103
Neural control
blood flow, 100–103, 136
cardiac, 122
cerebral, 119–120
cutaneous, 129–131
gastrointestinal, 124, 126
skeletal muscle, 112–116
Norepinephrine
adrenergic receptor stimulation, 111–112
release, 109
vasoactivity, 111–112

O

Opticoelectrical transducer, as blood velocity sensor, 160–169
Osmolarity, tissue vasodilatory effects, 105
Osmosis, in capillary filtration, 145–147

Osmotic pressure gradient, capillary filtration effect, 145–147
Oxygen, effect on blood flow, 103–105, 120, 121
Oxygen consumption, 127
Oxygen exchange, 143
Oxygenation, of blood, 76
Oxytocin, vasoactivity, 111

P

Parasympathetic nerve, 119–120, 126
Parenchyma, 12
Pentobarbital, 76, 79, 87
Perfusion, capillary, 117–119, 123, 126
Perfusion pressure
  autoregulatory response, 134, 135, 136, 137
  during hypertension, 133
  intestinal, 134–135
Peripheral vascular resistance, 102
Permeability, capillary, 76, 110, 143–144
Permeability-surface area product, 113, 114
pH, blood flow effects, 103
Phosphate, vasoactivity, 117
Photography
  blood flow velocity, 159–160
  heart, 90, 91
Photosensor, in blood velocity measurement, 160–169
Pia mater
  blood flow velocity, 188
  microscopic observation, 88–89
  vasculature, 55–57, 88–89
Platelet
  aggregation, 10, 66
  diameter, 66
Polypeptide, vasoactivity, 108
Pore, diffusion, 143–144, 150–151, 152
Portal vein, 47
Portal venule, 47
Postcapillary vascular resistance, 181
Postcapillary venule, 52, 141
Postcapillary vessel, 42, 63, 184, 190–191
Potassium
  effect on cerebral blood flow, 121
  vasoactivity, 105, 117
Precapillary sphincter, 15, 16, 17–18
  bat wing, 69
  capillary flow effects, 191
  cardiac, 31
  catecholamine sensitivity, 112
  contractility, 31, 61
  definition, 115–116
  diameter change, 190
  dilation, 102
  hepatic, 48
  intestinal, 37
  lack of innervation, 100
  pulmonary, 52
  spinotrapezius muscle, 29
  tenuissimus muscle, 27, 28
  vasomotion, 25, 31, 45, 102
Precapillary vascular resistance, 103–104, 181
Precapillary vessel, 12–13, 14–15, 16
  blood pressure, 27
  diameter, 184
  mesenteric, 42
  tenuissimus muscle, 27
Preferential channel, 25, 43, 69
  definition, 17
  mesenteric, 45–46
Pressure
  intraluminal, 106–107
  transmural, 107–108
Prostaglandin
  blood flow control, 117
  vasodilatory effects, 105–106, 130

Q

Quartz rod technique, 85–86, 87, 89, 90

R

Rabbit ear
  chamber, 83–84
  vasculature, 63–65
Rat, use in microvasculature studies, 77–78, 79, 87, 88, 89, 90–91
Renin, blood pressure effects, 110
Respiratory depression, 76
Resting tone, of blood vessels, 59
Reticular cell, 50

S

Sartorius muscle, 29–30
Secretin, vasoactivity, 126

Index

Serotonin, vasoactivity, 108, 109
Sinusoid, 47, 48
Skeletal muscle
   blood flow
      autoregulation, 131-132
      control, 112-119
      velocity profile, 188
   capillary perfusion, 117-119
   innervation, 112-116
Skin, blood flow, 129-131
Smooth muscle
   blood flow, 100-112
   contraction, 106-108
   myogenic response, 106-108
Solute exchange, 140-152
Solution chamber, 80
Spinotrapezius muscle, 29, 119
Spleen
   blood flow, 48-51
   microscopic observation, 87-88
Stomach
   arterial blood supply, 33, 34.-35
   exteriorization, 78-79
   microscopic observation, 78-79
   venous system, 35
Streak-image technique, 158, 166
Submucosa, gastric, 35
Surgical procedure, for microvasculature observation, 77-80, 87-90
Sweat gland, innervation, 130
Sympathetic cholinergic nerve, 116
Sympathetic nervous system, 100
Sympathetic vasoconstrictor nerve, 68, 112-113, 119-120, 122, 124, 126, 129-130, 136
Systole, 3, 55, 134

T

Television, closed circuit
   blood flow observation, 166-167
Temperature
   blood flow effects, 131
   effect on leukocyte number, 76
Tenuissimus muscle
   artery, 25
   blood flow, 119
   capillary network, 27, 28
   microvasculature, 25-29
   surgical preparation, 86-87

Tissue
   exteriorization, 76-82
   function, 69-70
   metabolism, 103
   microscopic observation, 75-94
Titanium chamber, 65-66,83
Transducer, blood pressure measurement, 170-174
Transendothelial channel, capillary, 149-150
Transillumination, 77, 84, 85-87, 89, 90, 91
Transmural pressure, 107-108, 116-117, 136
Transparent chamber, 63-65, 77, 82-85
Transport
   molecular, 140-152
   vesicular, 147-152

U

Urethane, 79, 86

V

Vascular bed, *see also* Capillary bed
   bat wing, 67-68
   cardiac, 133-135
   hamster cheek pouch, 80
   hemodynamics, 191-206
   mesenteric, 42, 46-47
   microscopic observation, 75-94
   pulmonary, 51-52
   structure, 191-206
   visualization, 75-94
Vascular capacitance, 124, 126
Vascular pattern, 191-206
Vascular pressure, *see* Blood pressure
Vascular reactivity, 55
Vascular resistance
   capillary, 115, 196, 197, 202-205, 208-211
   cardiac, 122
   cutaneous, 131
   gastrointestinal, 124, 126
   measurement, 186
   precapillary, 181
   postcapillary, 181
   sympathetic nerve control, 113-116
Vasoconstriction, *see also* Blood flow
   adrenergic receptor role, 111-112

Vasoconstriction (*continued*)
  angiotensin effects, 110
  arterial, 63-64, 68-69
  ateriole, 53, 54
  arteriovenous anastomosis, 64-65
  capillary pressure effects, 190
  cardiac, 122
  catecholamine effects, 111-112
  cerebral, 120
  hepatic, 48
  mechanism, 101-103
  myogenic response, 101-102
  neural control, 100
  oxygen effects, 104-105
  oxytocin effects, 111
  precapillary sphincter, 25, 31, 45
  temperature effects, 131
  vasopressin effects, 111
Vasoconstrictor, metabolic, 108-112
Vasoconstrictor nerve, 100-103, 112-113, 129-130, 136
Vasodilation
  adenine effects, 117
  adenosine effects, 105, 117
  adrenergic receptor role, 111-112
  anesthesia effects, 76
  arterial, 104-105
  arteriole, 43, 102
  carbon dioxide effects, 103, 117, 120, 121
  cardiac, 122
  catecholamine effects, 111-112
  histamine effects, 108, 109
  hormone effects, 126
  hydrogen ion concentration effects, 117
  kinin effects, 110
  magnesium effects, 117
  metabolic control, 117
  oxygen effects, 104-105
  phosphate effects, 117
  potassium effects, 105, 117
  precapillary sphincter, 102
  prostaglandins effects, 105-106
  serotonin effects, 108, 109
  spontaneous contractile activity, 101-102
  tissue osmolarity effects, 105
Vasodilator, metabolic, 108-112, 117
Vasodilator nerve, 61
Vasomotion, 70, *see also* Blood flow
  anesthesia effects, 76
  arterial, 29, 63
  arteriole, 25, 28, 38, 59
  capillary effects, 190, 191
  hormone effects, 126
  humoral factors, 108-112
  mesenteric, 47
  myogenic control, 106-108
  precapillary, 25, 27, 54
  venous, 29, 38, 69
Vasopressin, 111, 127
Vein, *see also* specific veins
  diameter, 67
  hemodynamics, 190
Velocity
  blood flow, 27, 54, 76, 157-169, 186-190, 207-208
  fluid, 207
Venous outflow, 37-38, 63
Venous plexus, submucosal, 35
Venous system
  cardiac, 30
  cremaster muscle, 24-25
  gastric, 35
  mesenteric, 39, 42
  tenuissimus muscle, 25
Ventricle, 3
Venule, 17
  blood flow, 37-38, 65
  collecting, 42
  diameter, 55, 67
  intestinal, 37-38
  portal, 47
  postcapillary, 17, 52, 141
  pressure function, 190
  tributaries, 56
Vesicle, capillary, 147-152
Vesicular transport, 147-152
Villi, intestinal, 38, 127-129, 135
Viscera, microvasculature, 33-53
Viscosity
  blood, 189, 207-208
  fluid, 207-208
Visual method, of blood flow measurement, 158
Visualization, of vascular bed, 75-94

W

Whytt, Robert, 10